BEET-ROOT SUGAR

AND

CULTIVATION OF THE BEET.

By E. B. GRANT.

BOSTON:
LEE AND SHEPARD.
1867.

STEREOTYPED AT THE
BOSTON STEREOTYPE FOUNDRY,
No. 4 Spring Lane.

GEO. C. RAND & AVERY, Cornhill Press.

PREFACE.

THE experience of Europe in the failure of their supply of cotton, caused by the late war, should teach the United States not to depend too exclusively upon foreign countries for her supply of so necessary an article as Sugar, of which the consumption is about 400,000 tons, while the production of all kinds within her borders is less than 50,000 tons; leaving 350,000 tons to be imported.

In case of war with a great maritime power, like England or France, which would, in part at least, prevent importation, sugars would necessarily advance enormously.

The emancipation of slaves in the only remaining strongholds of slavery — Cuba and Brazil (which is simply a question of a very few years) — would probably, at first, as it has always done elsewhere, diminish the production of sugar in those countries at least fifty per cent.

If such should prove to be the case, as this production exceeds 600,000 tons, the diminution would make so serious an inroad upon the ordinary supplies of the world that prices would materially advance.

It is, without doubt, within the power of the United

States to produce, within her own borders, not only all the sugar she requires for home consumption, but also to become a large sugar-exporting country, and that within a very few years. It is believed that the sugar beet is the plant destined to effect this revolution, and the reasons are set forth in the following pages.

The object of this book is to call attention to the importance of beet-sugar production in the Old World, and to demonstrate the advantage and feasibility of establishing it in the United States.

Part I. is chiefly devoted to the history of beet sugar, and the relative advantages of Europe and North America for its production.

Part II. is wholly agricultural in its character, and gives instruction for the choice of soil suitable for the cultivation of beets; the methods of preservation; of raising the seed; and of the preservation and use of the pulp.

Basset, in his work, "Guide Pratique du Fabricant de Sucre," says, "The manufacture of sugar from beets is one of the most important elements of public prosperity.

"Resting on agricultural progress and the wants of a constantly increasing population, allied by reason of the cattle which it supports with the production of meat and bread, based upon improving cultivation, it renders to modern society the greatest services, at the same time that it attains for itself the highest point of prosperity and glory to which any industry ever had the ambition to aspire."

CONTENTS.

PART I.

PAGE

SUGAR. Supply of the United States. — Manufacture of Beet-root Sugar in Europe. — Location favorable for its Production in the United States. — Value of Land and of Coal. 7

HISTORY OF BEET-ROOT SUGAR. Its Discovery. — Early Experiments. — Extent of its Production and Use in Europe. — Production in France and in Germany. — Prices from 1816 to 1866. — Consumption of Sugar in Europe. — Production of Sugar in the World. — Quality of Beet-root Sugar. — Taxes. 9–25

SUPPLY OF BEETS. Extent and Cost of their Cultivation. — Average Yield. — Saccharine Contents. — Profits of Cultivation. — Cost in France. — Probable Cost in the United States. — Advantages of Beet Culture. . . . 26–39

COST OF BEET-ROOT SUGAR IN FRANCE. Methods of Manufacture. — Detailed Cost of Manufacture. — Yield of Sugar. — Cost of Labor. 40–46

PROFITS ON BEET-ROOT SUGAR. Estimated Profits in the United States. — Ability to Compete with other Countries. — Recent Improvements in Europe. 47–54

PRODUCTION OF SUGAR IN VARIOUS COUNTRIES. Relative Yield of the Cane and of the Beet. 55, 56

ATTEMPTS TO MANUFACTURE BEET-ROOT SUGAR IN THE UNITED STATES. 57–63

GENERAL ADVANTAGES OF BEET-ROOT SUGAR MANUFACTURE. Effect produced in Europe by the Manufacture of Beet-root Sugar. — Probable Effect of its Introduction into the United States. — Opinions of Distinguished Men. 68–78

PART II.

THE BEET AND ITS CULTIVATION. Analysis of the Beet. — Varieties. — Choice of Beets. — Choice of Soil. — Climate. 79–92

METHOD OF RAISING THE SUGAR BEET. Preparing the Soil. —Ploughing. —Manuring. — Sowing. — Preparation of the Seed. — Seed Sowers. —Method of Sowing. —Weeding. — Thinning Out. — Cultivating. — Hoeing. — Transplanting. — Earthing Up. 92–106

HARVESTING. Signs of Maturity. — Effect of Frost. — Effect of Rains and of Drought. —Methods of Harvesting. 106–110

PRESERVATION OF BEETS. In Silos or Pits. — In Piles. — Method in Saxony. — In Massachusetts. 110–120

SEED. Importance of selecting the Best. — German Method of selecting Seed Beets. — Method of M. Vilmorin. — Soil and Manure suitable for Seed Beets. — Method of Preserving Beets for Seed. — Methods of Planting and subsequent Culture. —Time of Harvest. —Method of Saving the Seed. — Chinese Method of Cultivation. . 120–126

MANURES. General Effect of Manures on the Beet. — French and German Systems of Applying Manures. — Artificial Fertilizers. —Effect of different Fertilizers on Saccharine Contents of Beets. — Stassfurt Manures. — Analysis of the Ashes of Beets. 126–134

ROTATION OF CROPS. 134–136

BEET PULP. Method of Preserving Pulp. — Method of Feeding it to Cattle. 136–139

LEAVES OF BEETS. Their Uses. — Methods of Preservation. — Their Effect upon Milch Cows. 139–143

CULTURE OF THE BEET. Its Advantages to Farmers and the Country. 143

APPENDIX. 144–158

BEET-ROOT SUGAR.

PART I.

My attention was drawn to the question of the sugar supply of the United States by the very high prices prevailing in the spring of 1865, and I commenced the investigation of the subject early in the month of March of the same year.

Having satisfied myself of the firm basis of the beet-sugar industry in Europe, and that its establishment in the United States was not only practicable, but also promised to be highly remunerative, I spent several weeks in pursuit of that locality which possessed in the highest degree the advantages of cheap land, labor, fuel, transportation, and also a high market for sugar.

Most of these conditions are fulfilled in the region which I have selected, viz.,—the coal and prairie lands of Illinois, on the line of the Chicago and Rock Island Railroad,—the territory being intersected, not only by the above-named road, but also by the Illinois River and the Illinois Canal, which secure cheap, easy, and frequent transportation to Chicago, St. Louis, and the

Mississippi River, with all important points in every direction.

The condition sought that is unfulfilled is that of cheap labor; but it is a well-known fact that, notwithstanding the high price of labor, the peculiarly favorable nature of the soil of Illinois for the use of agricultural machinery enables the farmer of that state to cultivate land as cheaply as in any part of the world. For the enterprise in question, moreover, it is believed that the location is a favorable one, for it is in the midst of a German population, many of whom have had experience, in their own country, in the cultivation of beets and the manufacture of beet sugar.

Land can be bought at from twenty-five dollars to forty-five dollars per acre. "Slack" or coal screenings can be had at factory for one dollar and twenty-five cents per ton; and sugar is usually worth at Chicago, by reason of its distance from the present sources of supply, from one to one and one half cents per pound more than upon the seaboard.

Having satisfied myself that the manufacture of beet sugar in the State of Illinois promised to be profitable, it remained to acquaint myself with the process and condition of its manufacture in Europe; and for that purpose I sailed from the United States, early in December last, furnished with letters which have given me access to every establishment that I desired to visit in France and Germany.

HISTORY OF BEET SUGAR.

I will proceed to give a sketch of the history of the beet-sugar manufacture in Europe, but principally that of France, where this branch of industry is carried on more scientifically and successfully than in any other country.

The beet is supposed to be a native of Turkey, and to have been introduced into France about the year 1595.

In the year 1747, Margraff, a Prussian chemist, discovered that sugar was contained in beets, and advised his countrymen to cultivate them for the purpose of making sugar.

In 1773, Achard, a Prussian chemist, tried various experiments for the manufacture of beet sugar, under the patronage of Frederick the Great. Prevented for a time, by the death of Frederick, from pursuing his investigation, he did not again attempt it until 1795, when he planted sixty or seventy acres with beets. In the year 1799 he presented several loaves of beet sugar to the King of Prussia. He reported that he had produced a good quality of raw sugar at sixty-five centimes a kilogramme, or about six cents per pound, and expressed the belief that his process was susceptible of improvements that would reduce the cost one half. This report of Achard having been published in the annals of chemistry ("Annales de Chimie"), a committee, consisting of some of the most distinguished chemists of France, was appointed by the Institute, to investigate the merits of Achard's discovery. Their report stated that the amount of sugar extracted was rather less than one per cent., and the enterprise was abandoned, until Napoleon I. again called attention to the subject, and appointed a new committee to conduct further experiments. M. Deyeux, a member of this committee, made his report in 1810, and presented two loaves of sugar, equal in every respect to the best sugar from the cane.

In this report neither the percentage of sugar obtained nor the cost of production was given. Reports not well verified were published that in Germany from four to six per cent. of sugar had been obtained.

By the experiments of M. Barruel, from fifty to sixty per cent. only of juice was obtained from the beet; whereas the production at the present time is from eighty to eighty-five per cent. The yield of sugar was about one and one half per cent., while at the present time in France it is about seven; in Germany, eight to nine; and in Russia, nine to ten per cent. The cost was nearly thirty cents per pound, while at the present time it is about four cents.

M. Derosne, a Frenchman, obtained about this time two per cent. of sugar from the beet. Other experiments yielded two and one half per cent. A factory working 500 tons of beets in a season was considered quite extensive. There are establishments now in operation that work 60,000 tons.

A rasp then worked up about three tons per diem. Now, from 150 to 300 tons a day are consumed by one rasp.

In 1812 the continental blockade favored the establishment of the beet-sugar industry. The cost of manufacturing sugar was about 105 francs the hundred kilogrammes (say nine cents per pound).

Chemical schools and imperial factories were established, and government ordered the cultivation of 100,000 acres of beets. The sum of $200,000 was placed in the hands of the minister of agriculture, with which to encourage the production. Five hundred licenses to manufacture were given, and the indigenous sugar was exempt from duty for four years. The political events of 1814 caused the failure of all the beet-sugar manufacturers but one, M. Crespel Delisse, who continued to work.

In December, 1814, the impost on beet sugar was fixed at forty francs the one hundred kilogrammes, — about three and one third cents per pound, — and the duty on foreign sugars at fifty per cent. advance (say five cents per pound). This infused new life into the industry ; manufacturers introduced great improvements in their establishments, improving the processes of rasping and pressing to such an extent that they obtained seventy per cent. of juice from the beet, in lieu of fifty and sixty.

Bone black, or animal charcoal, was used in filtration. Machinery driven by wind and water, as well as by horses and oxen, replaced the more slow and costly processes of hand labor.

The yield of sugar was from three to four per cent., and of molasses about five per cent. M. Crespel Delisse claimed that he obtained five per cent. sugar and 4.8 molasses. The cost of manufacturing was about eighty-five francs the hundred kilogrammes (say seven cents per pound). From 1822 to 1830 the number of manufactories largely increased. The yield of sugar was about five per cent., and the cost of production from sixty to seventy francs the hundred kilogrammes (average, say five and one half cents per pound).

In 1825 France produced 5000 tons of sugar in over one hundred establishments. From 1830 to 1836 great progress was made. The sugar produced was of improved quality, and amounted to about five per cent. of the weight of beets worked. The introduction of steam power increased the means of production tenfold. In 1836 four hundred and thirty-six factories were in operation.

The intimate relation between this branch of industry and agriculture developed itself, and there were no longer unimproved lands in the vicinity of a sugar manufactory. In the department of the North, where the industry was most firmly established, the number of acres under cultivation in grain increased enormously, the beet pulp furnishing farmers with the means of feeding an increased number of cattle, thus providing the means of fertilizing an increased amount of land.

In 1837 government laid a manufacturer's tax on domestic sugars of fifteen francs the hundred kilogrammes (say one and one fourth cents per pound). This caused the failure of one hundred and sixty-six establishments.

In 1837 M. Payen, professor of the "School of Arts and Manufactures," in a communication to the "Royal and Central Agricultural Society," stated that beets in a fresh state contained ten per cent. of crystallizable sugar. They contained no uncrystallizable sugar, neither grape nor mannite.

Nevertheless, by the processes in actual use there was obtained but five or six per cent. in the two or three first months after harvest, and later in the season three to four per cent. only; the whole average being but four to five per cent.

He expressed the belief that inasmuch as the beet contained ninety-five per cent. of juice, while there was but seventy per cent. extracted, the yield of sugar might be largely increased.*

In order to stimulate improvement, the "Society for the Encouragement of Beet-sugar Manufacture" offered a prize of 10,000 francs to the person who should

* The sugar beet actually contains ninety-five per cent. of juice, of which only eighty is usually extracted, although eighty-five per cent. is occasionally obtained. Robert de Massy, of St. Quentin, in France, has invented a method by which he claims to obtain ninety-three per cent. The inventor is a very wealthy, as well as an ingenious man, and claims that his process will not only increase the yield of sugar from one to one and a half per cent., but will also materially lessen expenses, as it dispenses with all the hydraulic presses, hurdles, and sacks, besides diminishing the number of workmen required in the factory. — I visited Mr. De Massy's sugar factory at Busigny last winter with the proprietor, to see the apparatus in operation, but an accident prevented its working. Since my return, Mr. De Massy, through the "Journal des Fabricants de Sucre," invited all the manufacturers to visit Busigny on the 15th of May, for the purpose of seeing the apparatus at work. The amount of juice obtained at this trial was eighty-nine per cent.; but I infer from reading the article in which an account of the meeting is given, that the experiment did not thoroughly satisfy the manufacturers present of the value of the invention in its then existing state.

find the means of extracting from beets containing ten per cent. (without increasing the usual cost of manufacture) eight per cent. of crystallized sugar in the first four months of working, and 10,000 francs to any one who should extract eight per cent. crystallized sugar from any beets, without regard to the degree of richness or time of manufacture.

I annex a table in which is shown the contrast between the average results obtained in 1837, according to Mr. Payen, and those of 1865:—

	1837.	1865.
Yield of beets to an acre,	12 tons.	16 tons.
Price of beets per ton,	$3.00.	$3.25.
Percentage of sugar contained in beets,	10 per cent.	11 1-2 per cent.
Percentage of sugar produced from beets,	4 1-5 "	7 "
Cost of sugar per pound,	7 3-10 cents	4 cents
Sugar produced in France,	49,000 tons.	270,000 tons.

The production of sugar, which had risen to 49,000 tons in 1837, fell to 39,000 in 1839, and to 22,000 in 1840.

The history of the beet-sugar industry from the year 1837 nearly to the present day, is but the record of a struggle on the part of the cane-sugar manufacturers of the French colonies with their formidable rivals the beet-sugar manufacturers of France:—

In 1843,	the beet sugar produced, was,	28,000 tons.
"	colonial sugar imported,	83,000 "
In 1848,	beet sugar produced,	56,000 "
"	colonial sugar imported,	62,000 "
In 1850,	beet sugar produced,	64,000 "
"	colonial sugar imported,	46.000 "

To-day the French colonists have a protection on their sugars in France of five francs the hundred kilogrammes (about half a cent per pound). The beet-sugar manufacturers have no protection, competing at a disadvantage with French colonial sugar, and upon equal terms with the products of the rest of the world. Notwithstanding this, the following figures show the relative importance of the French traffic in native, foreign, and colonial sugars in the year 1865-6: —

Colonial importations, ...	76,103 tons.
Foreign "	144,083 "
Beet sugar manufactured, ...	270,000 "

Beet sugar fifty-five per cent. of the total traffic. The exportation of refined sugar for the same period was 114,150 tons, mostly of foreign and colonial sugars, owing to the policy of the French government, which, to encourage its commerce, accords an advantage to the refiner of foreign sugar for exportation, leaving the supply of the home consumption almost entirely in the hands of the manufacturers of beet sugar.

Except in the immediate vicinity of the seaboard cities of France, no sugar is used but the beet. The same is true of Germany. Not an ounce of any other is consumed in Paris, Vienna, Berlin, Dresden, Leipsic, or Munich.

In 1853-4 the high price of alcohol — one hundred and eighty-five francs the hectolitre (one dollar and fifty-seven cents per gallon) — induced some twenty manufacturers of sugar to convert their factories into distilleries, and in 1854-5 nearly one hundred more pursued the same course.

Since 1840 the production of beet sugar in France has doubled every ten years.

In 1830 the consumption of sugar in France was about two pounds per head, of which the beet-sugar manufacturer produced nine per cent.

In 1865 the consumption was over fourteen pounds per head, and the beet-sugar manufacturer produced more than enough to supply the home demand, although the total consumption had in the mean time increased tenfold.

There were sixty manufactories of beet sugar in Austria in 1840; in 1865 the number had increased to one hundred and forty.

The states of the Zollverein have nearly quadrupled their production in the past fifteen years, 52,586 tons having been produced in 1850, against 180,000 tons in 1865–6. In the same time the quantity of imported sugar has fallen from 52,568 tons to 12,562 tons, proving that cane sugar is almost entirely expelled from Germany. In 1865–6 there were thirty new establishments built in Germany, and many old ones enlarged their machinery.

It will be seen by the following table which gives an idea of the importance and progress of this industry, that although the number of factories was but ninety-six in 1845, against one hundred and forty-five in 1840, yet the amount of sugar produced was greater. Establishments were consolidated and enlarged, it being found economical to work upon a more extended scale. This table also shows the increasing tax paid upon the raw beets, which rose from twelve cents per ton in 1840 to three dollars and fifty-six cents in 1865.

TABLE

Showing approximately the Quantities of Beets used in the Manufactories of the Zollverein, the Products extracted, and Rate of Taxation, from 1840–1 *to* 1865–6.

Year.	Number of factories.	Beets used, tons.	Raw sugar, tons.	Molasses, tons.	Tons of beets to a ton of sugar.	Sugar, per cent.	Molasses, per cent.	Total yield, per cent.	Tax on roots per ton. Dol's.
1840	145	241,486	13,445	8,955	18	5.55	3.7	9.25	.12
1845	96	222,754	14,850	6,905	15	6.67	3.1	9.77	.71
1850	184	736,215	52,586	19,877	14	7.14	2.7	9.84	1.42
1851	234	914,495	60,966	27,434	15	6.67	3	9.67	1.42
1858	257	1,833,427	146,674	45,835	12.5	8	2.5	10.50	3.56
1860	247	1,467,701	126,526	35,224	11.6	8.62	2.4	11.02	3.56
1861	242	1,584,619	122,838	38,050	12.9	7.75	2.4	10.15	3.56
1862	247	1,835,663	138,042	44,055	13	7.52	2.4	9.92	3.56
1863	253	1,999,576	151,180	47,989	13.2	7.55	2.4	9.95	3.56
1864	270	2,079,729	165,978	49,913	12.4	7.98	2.4	10.28	3.56
1865	300	2,106,000	180,000	50,544	11.7	8.54	2.4	10.94	3.56

The average yield of sugar for the past eight years has been over eight per cent., and of molasses about 2.40; but this includes, of course, the results of all the poorly managed establishments, many of which are worked by the old process. I have visited several establishments where the yield of sugar averaged nine per cent., and of molasses two and a half to three per cent., throughout the season.

In France, too, where the whole average yield is perhaps rather less than seven per cent. of sugar, I know of establishments, working by improved processes, where the yield is from seven and three fourths to eight per cent. of superior sugar, and from three to three and a half per cent. of molasses.

There were 86,000 acres of land in cultivation with beets in France in 1850, and 297,000 acres in 1865. The product from this land was manufactured into sugar and alcohol, 270,000 tons of the former and 6,000,000 gallons of the latter having been produced in 1865.

The products obtained from beets in France, in about the following proportions, are, pulp, twenty per cent.; sugar, seven per cent.; alcohol, three fourths per cent.; potash, one fifth per cent.; soda, one tenth per cent.

The pulp is the refuse of the beet after the extraction of the juice. It is fed to cattle and sheep, which are extremely fond of it, and are quickly fattened upon it.

It is worth from two dollars and seventy-five cents to three dollars per ton at the factories, and is estimated to be worth, for feeding purposes, one third as much as the best hay.

After the sugar is extracted from the juice there remains about three per cent. of the original weight of the beet in the form of molasses, from which alcohol is distilled.

The molasses, which usually sells at from fifteen to eighteen cents per gallon, produces twenty-five per cent. of 90° alcohol. The cost of distillation is less than twenty cents per gallon of alcohol.

After the extraction of the alcohol, there remains

from ten to twelve per cent. of the weight of the molasses in the following salts, in the proportions given below: —

	Per cent.
Carbonate of potash, ...	40.33
Sulphate of potash, ...	2.46
Hydrochlorate of potash, ...	22.10
Soda, ...	34.14
Sulphur and divers matters, ...	.97
	100.00

The potash and the soda are extracted at a cost not exceeding three and a half cents per pound.

The following table shows the average prices, exclusive of duties, of No. 12 raw sugar in Paris from 1816 to 1828: —

1816	$12\frac{5}{10}$ cents.	1823	$8\frac{6}{10}$ cents.
1817	$11\frac{6}{10}$ "	1824	$10\frac{3}{10}$ "
1818	$12\frac{1}{10}$ "	1825	$9\frac{9}{10}$ "
1819	$11\frac{6}{10}$ "	1826	$10\frac{3}{10}$ "
1820	$10\frac{8}{10}$ "	1827	$9\frac{9}{10}$ "
1821	$10\frac{8}{10}$ "	1828	$9\frac{9}{10}$ "
1822	$7\frac{8}{10}$ "		

From 1828 to 1854 the price gradually fell, and the following table shows the average prices from 1854 to 1865: —

1854	$5\frac{8}{10}$ cents.	1860	$6\frac{1}{10}$ cents.
1855	6 "	1861	$5\frac{9}{10}$ "
1856	$6\frac{4}{10}$ "	1862	$5\frac{2}{10}$ "
1857	$7\frac{6}{10}$ "	1863	$5\frac{2}{10}$ "
1858	$5\frac{6}{10}$ "	1864	$5\frac{2}{10}$ "
1859	$6\frac{1}{10}$ "	1865	5 "

The price in April, 1866, was four and three fourths cents per pound.

The preceding table shows that the price of sugar has constanly fallen since 1816. Yet production has steadily increased.

It will be seen that the price of sugars, *exclusive of duties*, was in 1816 about three times greater than at present. But this does not fully convey an idea of the difference in the state of things existing then and now.

From 1816 to 1833 beet sugars were protected by a duty on foreign sugars varying from five to eight cents per pound.

From 1833 to 1840 they had a protection of two and one fourth to five and three fourth cents per pound.

From 1840 to 1860 they were protected by a duty of from one to three and a half cents per pound on foreign sugar.

From 1860 to the present time, not only has there been no protection as against foreign sugars, but sugars of the French colonies have had an advantage over all others of nearly half a cent per pound.

In addition to constantly diminishing price, with steadily decreasing protection, wages have doubled, and it is to increased skill alone that beet-sugar manufacture owes its present existence.

The following table shows the production of beet sugar in France from 1828 to 1865: —

Year.	Tons.	Year.	Tons.	Year.	Tons.
1828	4,665	1841	26,000	1854	77,000
1829	4,380	1842	30,000	1855	45,000
1830	5,500	1843	28,000	1856	92,000
1831	7,000	1844	30,000	1857	80,874
1832	9,000	1845	37,000	1858	150,444
1833	12,000	1846	49,000	1859	131,762
1834	20,000	1847	60,000	1860	130,000
1835	30,000	1848	56,000	1861	146,414
1836	40,000	1849	44,000	1862	173,675
1837	49,000	1850	64,000	1863	108,495
1838	47,000	1851	75,000	1864	145,745
1839	39,000	1852	60,000	1865	270,000
1840	22,000	1853	75,000		

The following table shows the number of beet-sugar manufactories in Europe, with their production, in 1865–6: —

France,	270,000 tons.	420 factories.		18 building.	
Holland,	5,000 "	8	"	2	"
Austria,	80,000 "	140	"	unknown.	
Zollverein,	180,000 "	300	"	14 building.	
Russia,	50,000 "	438	"	unknown.	
Belgium,	30,000 "	63	"	2 building.	
Poland,	14,000 "	54	"	2	"
Sweden,	1,000 "	3	"	unknown.	
Total,	630,000	1426		38	

It is stated by Mr. William Reed, an English authority, that Great Britain, which consumed 10,000 tons of sugar in 1700, consumes at the present time 566,000 tons.

Consumption of sugar in the following countries in 1865: —

	Tons.	Pounds per head.
Great Britain, ...	566,220	$37\frac{3}{10}$
France, ...	268,200	$14\frac{1}{9}$
Switzerland, ...	18,000	$14\frac{2}{5}$
Zollverein, ...	150,000	9
Spain and Portugal, . .	60,000	6
Italy, Turkey, and Greece,	110,000	$5\frac{1}{2}$
Sweden and Norway, . .	15,000	5
Poland, ...	10,000	4
Austria, ...	50,000	$2\frac{7}{9}$
Russia, ...	57,000	$1\frac{2}{3}$
Holland, Belgium, &c.,	50,000	
Total, ...	1,354,420	

From the two preceding tables it appears that Europe produced from beets in 1865–6 nearly one half her consumption.

The chief sugar-producing plants are the sugar-cane (*Arundo saccharifera*), the beet (*Beta vulgaris*), the date-palm (*Phœnix sylvestris*), and the sugar-maple (*Acer saccharinum*).

The total production of sugar in the world is not far from 2,800,000 tons, in about the following proportions:—

Sugar-cane,	71.42	per cent., or	2,000,000	tons.
Beet,	22.50	" "	630,000	"
Palm,	5.00	" "	140,000	"
Maple,	1.08	" "	30,000	"
			2,800,000	

It will be seen that the beet furnishes nearly one

quarter (twenty-two and a half per cent.) of the sugar of the world.

Arnold Baruchson & Co., in giving the statement of the London sugar market in their circular of March 10, 1866, say, "The greatest attention ought to be paid by dealers to the beet-sugar crop of Europe, for it is clear that before long she will produce all her own sugar."

There was formerly a prejudice in the minds of many people against beet sugar; but it is perfectly well ascertained, that, if properly refined, it cannot be distinguished from the best sugar of sugar-cane, either by taste, appearance, or chemical analysis: the two are identical.

William Reed, of London, says, in his recently published work, "History of Sugar and Sugar-yielding Plants," "Beet-root sugar is not only identical in every respect with cane sugar, but much of the Dutch lump sugar is actually the produce of beet root. The circumstance cannot be too much insisted upon, that the seeming distinction between yellow beet sugar and yellow cane sugar depends on the extraneous colored matters present. These, when eliminated by refining, leave white materials in all respects identical. There is positively no difference between these two, whether of color or of grain. Grain or crystals can from either be developed to the size of the largest candy if desired; in fact, at the present moment (1866), France is sending here large white crystals, produced from beet root, to compete with London, Bristol, and Scotch, and other crystal manufactories."

With the exception of London, most of the principal

cities of Europe use no other sugar than that of the beet; and even in England the consumption is rapidly increasing, Great Britain having, in the year 1865, imported 70,000 tons, which is in high favor with the refiners.

The "Journal des Fabricants de Sucre," in its issue of January 4, 1866, says, "One of the most remarkable and interesting facts of the past year is the exportation of considerable quantities of beet sugar from France to England — a country that not many years ago tried to stifle the beet-sugar industry in its infancy."

Referring to the fact that Achard, the Prusian chemist, stated that, after the first report of his discoveries in making sugar from the beet had been published, the English government, frightened by the effect it might have upon trade with their West India colonies, offered him a large sum of money to acknowledge publicly that he had been mistaken in the result of his experiments. But he indignantly refused the humiliating offer, and continued to publish the results of his labors.

The cost of producing from the beet a pure white sugar, entirely free from unpleasant smell or taste, is but a trifle more than is required to produce a lower grade. In Germany refined loaf sugar is produced directly from the beet. In France the brown is first produced, and then refined. Within the last two years, however, sugar has been produced of such purity and whiteness, that it has been sold directly for consumption without refining; and there is no question that the peculiar odor of the beet may be entirely got rid of in the manufactory.

Such is the present condition of beet-sugar manufacture in Europe. More than one third of the sugar there consumed is made from beets; and the progress of the industry is such, that it is perfectly clear, that within a few years the importation of sugar into Europe will entirely cease.

It is the constant effort of the French sugar manufacturer at the present day to induce government to reduce the duties and imposts on sugar, feeling that the reduction in the price consequent upon such action would largely increase consumption. He does not ask for protection against the manufacturers of cane sugar in any part of the world; for although the industry is entirely the creation of the protective policy, yet under it so great an amount of skill has been acquired, and the cost of manufacture has consequently been so reduced, that he is now able to compete upon equal terms with the whole world.

In France, the impost is laid upon the sugar produced; in Belgium, it was formerly laid upon the juice expressed from the beets; but at present it is upon the sugar, as in France; in Germany, upon the beets; in Austria, upon the sugar produced, or upon an agreed estimate of the capacity of the mill; in Russia, upon the hydraulic presses. It varies in the different countries from forty to eighty-five dollars per ton.

SUPPLY OF BEETS.

Having given an account of the rise and progress of the sugar industry in Europe, and demonstrated, as I trust, that it rests upon a firm basis, I shall proceed to consider the feasibility of establishing it in this country.

In comparing the relative positions of the two countries, I shall draw my comparisons chiefly with France, as the representative of Europe, the conditions of trade there being more nearly akin to those of the United States than in any other country; reliable statistics in this department of industry are more readily procured there than elsewhere in Europe, and the spirit of enterprise is so great among Frenchmen, that whatever improvement in the manufacture of sugar has been originated elsewhere, it has been seized upon, improved, and perfected in France.

And first as to the ability to procure in the United States raw beets, of good saccharine properties, upon reasonable terms.

The experience of Europe shows that beet of rich quality can be profitably cultivated from the Mediterranean to the North Sea, and from the Atlantic to the heart of Russia.

M. Mauny de Mornay says of the beet, that "all climates seem to suit it. It flourishes in the north and in the south. Moisture favors its development, but drought does not prevent its yielding good products. It may be regarded as the only root cultivated in Provence that also succeeds in the centre of the empire." Tomlinson says, in his Cyclopædia, "It has been shown by practical experiment and chemical analysis, that there is no material difference in beet grown over a region extending from the Atlantic to the Caspian Sea, and from the Mediterranean Sea nearly to the Arctic Ocean."

The universal testimony of the chemists, manufacturers, and farmers, with whom I conversed, was, that any good wheat land was suitable for beets. The sugar beet is almost identical with the mangel wurzel, the cultivation of which for stock has been very extensively and successfully practised in the Northern and Western States.

Repeated analyses made in the United States of beets, as well as of carrots, and other sugar-containing vegetables, show that they contain as much sugar as similar vegetables in Europe.

An analysis made of sugar beets, raised in Illinois, showed that they contained twelve and one half per cent. (12½) of crystallizable sugar in October, and eleven and four tenths per cent. in the following spring. A fair average percentage of sugar in the beet of France is eleven and one half per cent., in Germany it is about thirteen per cent., and in Russia even richer.

The quality of the beet has been very much improved within a few years, and within the last year extraordinary results have been attained, beets having been produced, containing even as high as eighteen per cent. of sugar. In one instance twenty-one per cent. was contained.

The quality of the beet, as well as the amount extracted from it, is largely affected by legislation and the price of labor. In France the impost tax is laid on the sugar produced; the consequence is, that the farmer strives for large crops, beets being sold by the ton, and he pays comparatively little heed to the quality.

In Germany, however, the impost is laid upon the beet; the cultivator consequently strives to produce a beet rich in sugar, paying greater attention to quality than to quantity.

In France labor is comparatively high, and the manufacturer is contented to obtain in sugar and molasses within two per cent. of all the saccharine matter contained in the beet, the extraction of the last two per cent. being costly in labor. The remaining pulp is also better for cattle than when a greater proportion is extracted.

In Germany, where wages are low, the pulp is more completely exhausted, and the manufacturer is not satisfied unless he obtains, in sugar and molasses, within from one half to one per cent. of all the existing saccharine matter.

A crop of beets was raised in Illinois, two years ago last summer, under the following disadvantageous circumstances. New prairie land was broken up, and the seed planted on the upturned sod — a course rarely pursued by good farmers anywhere; the beet requiring for its proper development a soil previously cultivated, in which the sod has been entirely rotted. The season was extremely dry, and the yield averaged from ten to twelve tons only, to the acre, of beets containing about twelve per cent. in sugar. The total cost, including the breaking up of the land, harvesting, and transportation, was three dollars and forty cents per ton.

In France the average yield of beets is from fifteen to eighteen tons per acre, frequently rising to thirty, and often to forty tons, while in one instance within my knowledge, nearly sixty-two tons were produced from a single acre. There is also an authentic account of a crop of over sixty-eight tons to an acre. In 1865 whole districts produced thirty-two tons per acre.

The cost of producing an acre of beets in Illinois, where all the conditions favor cheap cultivation, would not much exceed the cost of a crop of sorghum, which is estimated as low as thirty-five dollars, and is certainly not more than forty-five dollars per acre.

According to Flint's "Agriculture of Massachusetts," F. H. Williams, of Sunderland, cultivated one hundred and eighty-four rods, or an acre and an eighth, of land in broom-corn, at a cost of $38.32. This, including harvesting, cleaning the seed, and also eighteen dollars for manure, makes a total cost per acre of less that $34.

The same authority states that Alonzo P. Goodridge, of Worcester North, cultivated a crop of ruta bagas at a cost of $70 per acre, including $32 worth of manure. Yield, 43,880 pounds, or more than 19½ tons, to an acre. Cost, $3.59 per ton.

Mr. Goodridge also raised a crop of sugar beets at the same cost, and with the same amount and value of manure. Yield, 38,520 pounds, or about 17¼ tons, to an acre. Cost, $4.05 per ton.

S. D. Smith, of West Springfield, raised a crop of

sugar beets at a cost, of $38 per acre, including $16 for manure. Yield, 38,070 pounds, or 17 tons, to an acre. Cost, $2.23 per ton.

William Birnie, of Springfield, raised a crop of mangel-wurzel, in 1859, on 2½ acres of land, at a cost of $82 per acre, including $40 per acre for manure. Yield of mangel-wurzel, 76,000 pounds, or nearly 34 tons per acre. There were also harvested on the same land 400 heads of cabbage, besides 30 two-horse loads of beet-tops for milch cows. Cost per ton of beets, excluding value of tops or of cabbages, $2.38.

Mr. Birnie says, "I estimate that the improved condition of the land, after the crop is taken off, will more than balance the interest on its cost for the year."

Dr. Long, of Holyoke, raised a crop of ruta bagas, in 1860, at a cost of $48 per acre, including $12 for manure. Yield, 43,608 pounds, or nearly 20 tons, per acre. Cost, $2.40 per ton.

W. G. Wyman, of Worcester North, raised a crop of ruta bagas, at a cost of $50 per acre, including $36 for manure. Yield, 49,600 pounds, or more than 22 tons, per acre. Cost, $2.27 per ton.

According to the United States Agricultural Report for 1864, Thomas Messinger, of Long Island, N. Y., raised a crop of yellow globe mangel-wurzel at a cost of $57 per acre, including rent and every other expense. Yield, 111,000 pounds, or more than 49½ tons, to an acre. Cost, $1.15 per ton.

Tabular Statement of the Crops described.

Name of cultivators.	Crops.	Yield per ac., tons.	Cost of manure per acre.	Total cost per acre.	Cost per ton.
Alonzo Flint,	Broom-corn,	——	$16	$34	$——
A. P. Goodridge,	Ruta baga,	19½	32	70	3.59
Dr. Long,	Ruta baga,	20	12	48	2.40
W. G. Wyman,	Ruta baga,	22	36	50	2.27
Wm. Birnie,	Mangel-wurzel,	34	40	82	2.38
Thos. Messinger,	Mangel-wurzel,	49½	—	57	1.15
A. P. Goodridge,	Sugar beet,	17¼	32	70	4.05
S. D. Smith,	Sugar beet,	17	16	38	2.23
	Average	25.6	26.28	56.12	2.72

The average yield of roots to an acre was 25$\frac{6}{10}$ tons; the cost per ton was $2.72; the value of manure applied was $26.28; and the average gross cost of cultivation was $56.12 per acre.

The cost of cultivation, exclusive of manure, was $29.84 per acre, or $1.16 per ton of roots.

The usual average cost of cultivating sorghum, broom-corn, mangel-wurzel, and sugar beets is about the same.

The average price paid for beets in France, in 1865, was eighteen francs, say three dollars and fifty cents per ton; but at the close of the season, some were bought as low as two dollars per ton.

The average price for the last twenty years has been probably about three dollars and twelve cents per ton.

An acre of land producing twenty tons of beets, sold at three dollars and fifty cents per ton, would yield seventy dollars, — and with a yield of thirty tons one hundred and five dollars per acre.

What other crop could an Illinois farmer cultivate

that would yield him such a return? The following table shows that the principal crops raised in the Northern and Western States do not yield anything like such returns.

TABLE

Showing the Average Yield and Cash Value of Corn, Wheat, Rye, and Oats, on one acre of land, in twenty-two of the United States, for four years, from 1862 *to* 1865 *inclusive, according to the Report of the Agricultural Department for June,* 1866.

	Bushels.		Price per bushel.	Value per acre.
Corn, . .	32.99	per acre.	$ .86	$28.57
Wheat, .	14.34	"	1.57	22.44
Rye, . .	15.94	"	1.03	15.98
Oats, . .	28.56	"	.58	16.52

Average value of crops, per acre, $20.87.

The introduction of the manufacture of beet sugar in the West would give to the farmer a market for beets at his own door, and the establishment of a manufacturing population in his vicinity would give him a home market for the other productions of his farm.

In France the manufacturer contracts with the farmer for the culture of a certain number of acres in beets, at a fixed price per ton, and the crop is always sold in advance of its production.

The relative cost, in the department of the Maine et Loire, of raising an acre of beets, and an acre of wheat, by the same cultivator, and in the same year, is shown by the following figures. It is fair to remark, however, that labor in the region referred to is somewhat lower than in the north of France, where the beet is most extensively cultivated.

The total cost of cultivating and harvesting the beets on 580$\frac{8}{10}$ acres of land was as follows: —

Four Ploughings,	$9.18	per acre . .	$5,335.34
Manures, . . .	9.77	" " . .	5,676.31
Seeds, ...	.53	" " . .	310.46
Sowing, ...	1.84	" " . .	1,078.35
Cultivation, . . .	3.56	" " . .	2,069.10
Harvesting, . . .	1.42	" " . .	827.64
Transportation, .	1.18	" " . .	690.09
Sundries, . . .	.27	" " . .	156.26
Total, . . .	$27.75	" " .	$16,143.55

The total cost of cultivating and harvesting the wheat on 511$\frac{2}{10}$ acres of land was as follows: —

Ploughings, ...	$4.04	per acre	$2,065.37
Manures, ...	7.46	" " .	3,817.68
Seed-sowing, ...	3.55	" " .	1,818.30
Harrowing and rolling,	1.28	" " .	658.98
Harvesting and threshing,	3.40	" " .	1,745.12
Sundries, ...	.27	" " .	138.81
Total, ...	$20.00	" "	$10,244.26

From the above figures it appears that the cost of cultivating and harvesting an acre of beets was $27.75, and of an acre of wheat $20.00. Rent of land is not included in either account. The cost, then, of the acre of beets, was nearly thirty-eight per cent. more than that of the acre of wheat.

The cost of preparing and planting the ground in Illinois with a crop of beets would not exceed that of preparing and planting it with corn, for it would all be done by the same machinery that is now used. The increase of cost would arise from the greater amount of hand labor required on the beets to keep them entirely free from weeds. In France this labor is all done by the piece. The following are the prices paid for each operation subsequent to planting the seed upon the above-described field, containing $580\frac{8}{10}$ acres:—

First weeding, ...	$1.18	per acre.
Second weeding, ...	1.03	"
Third weeding, ...	.90	"
Thinning out, ...	.23	"
Pulling the beets, ...	1.42	"
Loading into wagon, . . .	.03	per ton.
Putting into "silos," . . .	.04	"

At these prices the workmen make from thirty-eight to forty-two cents per day. Much of the work is done by women and children.

On a crop of twenty tons to the acre, the cost of this labor would amount to $6.16 per acre. It is certainly safe to assume that the same work would not cost over twenty dollars per acre in this country; for I have found that the prices of labor in the United States are certainly not more than three times those prevailing in France, where a farm hand gets from fifty to sixty cents per day in gold.

The usually estimated cost of cultivating beets in France is from four hundred and fifty to six hundred francs per hectare, which is from thirty-five to forty-eight dollars per acre. This includes taxes, and also rent of land, which latter varies from eight to twenty-five dollars per acre per annum; and manures, which are applied at a cost of from ten to fifteen dollars per acre. Labor, of men, horses, and oxen, including ploughing, harvesting, and transport of crop to the manufactory, does not materially exceed fifteen dollars per acre.

I submit here the estimate of a practical French gentleman upon the cost of labor on an acre of beets.

Ploughing, ...	$5.54
Weeding, ...	3.96
Harvesting, ...	1.98
Transport, ...	3.96
Total, ...	$15.44

I can see no reason, then, why the western farmer cannot cultivate an acre of beets at a cost certainly not exceeding forty-five to fifty dollars, for the COST of his acre of land will not average TWICE THE ANNUAL RENT of the acre in France; and unless the present system of cultivation is materially changed, he will not apply fifteen dollars worth of manure to the acre, as they do in France. The use of labor-saving machines would probably enable him to diminish considerably the amount of hand labor employed, as compared with France. Even if he employ the same amount, and pay three times the prices paid by the French, not only for his laborers, but for his teams also, his work will not cost him over forty-five dollars per acre.

Assuming that the cost of cultivating an acre of beets would be even as high as sixty dollars per acre, — which is from fifteen to twenty-five dollars more than the cost of an acre of sorghum, — that the crop produced would be as great as that of a fair yield in France, or say twenty tons, then at four dollars per ton the crop would produce eighty dollars, leaving a *direct* net profit of twenty dollars per acre — a sum nearly as great as the gross receipts average at present, as shown by table on page 32.

I have said a *direct* net profit of twenty dollars per acre, because it has been found in Europe that there is also an *indirect* profit on the beet crop in the large increase of crops succeeding it, and in the cattle supported upon the pulp; experiments having conclusively proved that lands now yield from two to three times as much grain, and support from eight to ten times as many cattle, in the beet-growing districts as they did before the beet was introduced. The great beet-producing districts of France are the grain districts, and cattle districts also. The three branches of agriculture always co-exist.

David Lee Child published, in 1840, a book, to which further reference will be made hereafter. He cultivated sugar beets in Northampton, in this state, in 1838–9. He stated, as the result of his observation in France in 1836, that "the crops of beets in that country averaged about thirteen tons to the acre," * and that the result at Northampton was about the same. The

* Since Mr. Child's visit, cultivation has not only largely increased the production per acre, but it has considerably improved the saccharine properties of the beet.

sugar contained in the French beet was ten to ten and one half per cent., and in those raised at Northampton seven and one half to nine per cent. He attributed "the inferiority in richness to the inexperience of cultivators, and mainly to improper manuring. The probability is, that with equal culture our beets will surpass, in saccharine richness, those of France."

Mr. Child estimated the cost of raising a crop of beets at forty-two dollars per acre. He "had seen a great number of estimates based on more or less practice; and the great agreement which we find among them satisfies us that the general result may be relied upon. They are all very near forty dollars per acre. The lowest is thirty-five dollars and the highest is forty-four dollars."

At the same time Mr. Child estimated the cost of cultivating an acre of corn at thirty-one dollars and fifty cents, and an acre of broom corn at forty-two dollars. He says that the cost of cultivating an acre of beets and that of an acre of broom corn are exactly alike.

This corresponds with what I have said about the sorghum, the cultivation of which is identical with that of broom corn. He says, moreover, in reference to the corn and broom-corn crops, —

"But neither of these crops is an enriching or a cleaning crop: the beet is both, exterminating every noxious plant, and leaving good stuff on the ground, which ploughed in is equal to a quarter or half manuring, i. e., to five or ten loads of manure per acre and the expense of carting it."

In cane-sugar-producing countries the number of acres "tended" by a hand varies from one to five, according as agricultural machinery is more or less used. The cane in Louisiana is an eight or nine months' crop, and is cut before maturity.

In the West Indies it is in cultivation, before cutting, for a period of from eleven to fifteen months. The beet grows to maturity in France in from four to five months; in the United States in from three to four months. In France, with the aid of a horse, one hand will easily "tend" five acres of beets. I know of instances where a hand, with a horse, has done the whole work on five hectares, or twelve acres, of beets.

Mr. Child, in 1839, estimated that the whole number of days' labor on an acre of beets would vary from fifteen to nineteen.

In Illinois, a man, with a pair of horses, tends easily fifty acres of corn, and far more than that amount has been cultivated by one hand. I claim, therefore, that with the improved methods of cultivation now in practice, a man can easily cultivate six acres of beets in four months, and have more than half his time for other labors. The cultivation of six acres of cane would occupy a man exclusively for eight months. The labor, then, upon the acre of cane, is, at least, twice that on an acre of beets.

It will be shown that the product, per acre, of sugar from beets, is greater than the general average from cane.

But the advantages in favor of beet culture do not stop here. The cane crop is exhausting; it is a bad forerunner of other crops; the ground on which it is cultivated must lie fallow at least half the time; it feeds and fattens no sheep, cattle, nor swine; consequently, it affords little material for enriching the soil. The beet, on the contrary, is an enriching and cleaning crop. It requires no fallow; it is the very best known forerunner of other crops; it feeds multitudes of stock, and, instead of impoverishing the soil, constantly improves it.

In fact, there can be no doubt that the beet crop will be found to be as profitable to the farmer here as it unquestionably has been to the European farmer. The farmers of the west possess many great advantages over those of Europe.

They have a virgin soil prodigiously productive, easily cultivated, and of low cost, and agricultural machinery with which one man will do the work of a dozen. Probably, notwithstanding the high price of labor, there is no other country in which an acre of land is cultivated so cheaply as in the west.

I have conversed with a great many farmers in no less than twelve of the Northern and Western States, and have found no one who did not say that there would be no difficulty in getting all the beets we could consume for less than four dollars per ton. The impression among those farmers generally was, that it would cost from forty to fifty dollars an acre to raise a crop of beets; some placed it as low as thirty-five, and none over fifty dollars. If these estimates should prove to be correct, the cost of beets, with an ordinary yield, would be from two to three dollars per ton.

If it be true, then, that beets equally rich in sugar can be raised in the west as cheaply as in Europe, it only remains to inquire if that sugar can be extracted at a profit.

COST OF BEET SUGAR IN FRANCE.

There are various methods of making sugar from beets employed in Europe, of which the following are but a part: —

The old method of rasping, pressing, treating with lime, evaporating in open boilers, crystallizing in large moulds or in pans, draining, and crushing.

This method, in some factories, is modified by the introduction of the vacuum pan. In others the centrifugal machine takes the place of the slower method of moulds and of pans, for the purpose of throwing off the molasses.

In other establishments, instead of using hydraulic presses, juice is extracted from the pulp in centrifugal machines in which large quantities of water are used.

In others the "process of diffusion," so called, by which the beets are cut into thin slices, and the saccharine matter exhausted by steeping them in water in a series of vessels.

In others the process of "maceration" is applied to small slices of beets, called "cossettes," which are dried and then steeped in water in a range of "macerators."

In others there is a single saturation with carbonic acid gas after defecation.

In others the "Maumené process," or the system of cold defecation, is employed.

In others the sirup of the beets is "strengthened"

by the addition of sugar, and the refined loaf is produced directly from the beet.

In some establishments the old-fashioned "scum press," worked by hand, is seen, while others have "hydraulic scum presses." A score of different methods are employed in various parts of Europe for the treatment of the "scum."

In my judgment, however, incomparably the best process is the system of "double carbonitation," so called, of Perier and Possoz.

This method reduces the quantity of bone black required to a very small amount, allowing the beets to be worked later in the spring, producing a larger percentage of sugar, of better quality and at lower cost, than by any other method.

Taken in conjunction with the "hydraulic press," "Riedel's filter press," for the treatment of scums, the "carbonitation trouble," and, possibly, the "Joly rasp," it leaves little to be desired, and is the one that I heartily recommend for adoption.

In France the expense of manufacturing raw sugar, including the cost of the beets, varies from three to four cents per pound.

The average expenses of converting 1,000 tons of beets into sugar by the best processes are about as follows, not including taxes or interest on capital: —

1,000 tons beets @ $3.80, ...	$3,800
Coal, 120 tons, @ $3.00, ...	360
Bone-black waste, ...	300
Sacks for pulp, 250, @ 70 cts., ...	175
Labor, 220 men 5 days @ 70 cts., . .	770

Administration and salaries, ...		200
Lighting, ...		50
General expenses, insurance, ...		250
Lime, metals, rasp blades, repairs, &c.,		845
		6,750
From this is to be deducted, say		
200 tons pulp @ $2.50, ...	500	
30 " molasses @ $.22, . . .	660 =	1,160
Leaving, as total cost of working 1,000 tons beets, ...		$5,590

The cost per pound of sugar produced varies in accordance with the percentage of yield, as shown in the following table :—

Yield.	Sugar.	Cost per pound.
6 per cent.	134,440 lbs.	4.15 cts.
7 "	156,800 "	3.56 "
8 "	179,200 "	3.10 "

In one establishment that I visited in France, I asked in writing of the proprietor, to whom I had letters that warranted me in doing so, his percentage of sugar and molasses, and the cost of manufacturing.

This gentleman had been very successful, kept his accounts with great accuracy, and, as he manufactured by the old process, I selected him as a good representative of the old system, and asked him many questions, which he answered with great courtesy and in the fullest and most satisfactory manner. His yield of juice was eighty per cent. of the beets worked; his percentage of sugar was 6.85, and of molasses 2.75 per cent. *of the juice.*

This gives a result of 5.48 per cent. of sugar and 2.2 per cent. molasses on the beets worked, which was the poorest result with which I met.

In reply to my question as to the expense of converting a ton of beets into sugar, I shall give a literal translation of his reply, stating that the estimate was made from the business of nine years, in which time he had made improvements and enlargements of his mill, all of which were charged to expenses: —

"Hand labor, general expenses, ten per cent. depreciation of machinery, coal, taxes, in one word, *every* expense, even those for enlargements of works and improvements of machinery, amount to 13.75 francs the 1,000 kilogrammes of beets."

This is about $2.60 per ton of beets worked. The average price paid for beets in the above-described establishment was eighteen francs the 1,000 kilogrammes, or $3.42 per ton, making the total cost of a ton of beets and its conversion into sugar $6.02. From this is to be deducted the value of the pulp and molasses: —

Say, for 1,000 tons of beets @ $3.42, ...	$3,420	
Manufacturing 1,000 tons of beets @ $2.60, ...	2,600 =	$6,020
Less, 200 tons pulp @ $2.50, . .	500	
22 " molasses @ $22, .	484 =	984
		$5,036

Yield of sugar at 5.48 per cent., 54.8 tons, or 122,752 pounds, leaving the net cost of a pound of sugar 4$\frac{1}{10}$ cents.

The expense for labor at 3½ francs, or sixty-six cents, per day (the average) was ninety-two cents per ton of beets worked, being thirty-five per cent. of the cost of converting a ton of beets into sugar, and 15.2 per cent. of the *total cost*, including the price paid for the beets. This, if charged entirely to sugar, would make the cost of labor in a pound of sugar six mills.

Inquiry has satisfied me that the expense of manufacturing 1,000 kilogrammes, or 2,200 pounds, of beets into sugar in France, including in the expenses taxes, interest on capital, and depreciation of machinery, averages from eighteen to twenty francs, or $3.47 to $3.87 per ton of beets. In some cases it is as low as fifteen francs, or $2.88, per ton, and in others as high as twenty-two francs, or $4.25, per ton. In the case quoted above it was 13.75 francs, or $2.60, per ton.

The expense for labor in the best establishments is, as a rule, about twenty-five per cent. of the cost of *manufacturing*.

From these figures, which I know to be reliable, the cost of a pound of sugar and the proportion due to labor are shown in the following table; labor being reckoned at sixty-six cents per day and the cost of beets at $3.80 per ton; yield of molasses at two and one half per cent., price $22 per ton; pulp twenty per cent., price $2.50 per ton.

Cost of Labor and Total Cost per Pound of converting Beets into Sugar.

Manufacturing cost per ton of beet.	Yield.	Cost of labor per pound.	Total cost per pound.
$2.88	6 per cent.	$5\frac{3}{10}$ mills.	$4\frac{1}{10}$ cents.
	7 "	$4\frac{5}{10}$ "	$3\frac{6}{10}$ "
	8 "	4 "	$3\frac{1}{10}$ "
3.47	6 "	$6\frac{4}{10}$ "	$4\frac{6}{10}$ "
	7 "	$5\frac{5}{10}$ "	$3\frac{9}{10}$ "
	8 "	$4\frac{7}{10}$ "	$3\frac{4}{10}$ "
3.87	6 "	$7\frac{1}{10}$ "	$4\frac{9}{10}$ "
	7 "	$6\frac{1}{10}$ "	$4\frac{2}{10}$ "
	8 "	$5\frac{4}{10}$ "	$3\frac{6}{10}$ "
4.25	6 "	$7\frac{9}{10}$ "	$5\frac{2}{10}$ "
	7 "	$6\frac{7}{10}$ "	$4\frac{4}{10}$ "
	8 "	$5\frac{9}{10}$ "	$3\frac{9}{10}$ "

I know of an establishment in France where the total cost of producing sugar, exclusive of interest on capital, is but thirty-six francs per 1,000 kilogrammes of beets, or $3\frac{1}{10}$ cents per pound of sugar.

The yield of sugar is about eight per cent., of which four and one half per cent. is of a quality fit for direct consumption, and would bring fifteen cents per pound here to-day. Two and one half per cent. is of a grade better than No. 14, and one per cent. is equal to No. 12. In another about the same amount and quality is produced at a cost of $3\frac{7}{10}$ cents per pound.

I know of another establishment where the total cost, including every expense, interest on capital at five per cent., and depreciation of machinery at ten per cent., was in 1865–6 but the fraction of a mill over four cents per pound.

The amount of sugar produced was seven and one half per cent.; but the quality was not so good as in the previously described cases, although the first quality, which amounted to four per cent. of the beets worked, sold readily at seventy-five francs the hundred kilogrammes, or six and one half cents per pound.

PROFITS ON BEET SUGAR.

It is believed that the only material item of expense in the manufacture of sugar that would be greater in the United States than in France is the single one of labor. All others in excess of those of France are here more than offset by the lower cost of coal, of land, and of taxation.

In relation to labor it is well known that in the United States the use of labor-saving machines is greater than in any other country, because the high price of labor has stimulated their invention. It is a fact that the number of hands employed in sugar *refineries* in this country is much smaller than in European establishments of the same capacity of production, and it would doubtless be possible to effect some saving in that direction as compared with France in an American sugar *manufactory*.

The labor in a beet-sugar factory in this country would certainly not require a greater number of men than is required in a similar establishment in France. But, assuming that the same number would be necessary, it is proper to ascertain the exact relation that the price of labor bears to the cost of production.

In Europe the number of skilled hands required in a sugar manufactory is very small, the great proportion of workmen being common farm laborers, who work in the fields in summer and in the mills in winter. The making of beet sugar is only carried on in the fall and winter months, say from October to

February. With us, by reason of a more favorable climate, not only for the earlier development, but also for the better preservation of the beet, it could be extended from September to March, or even later. It will be acknowledged that these are the months in which labor in this country can be most readily and reasonably procured. The probability is, inasmuch as the establishment of this industry in Illinois would permit the hiring of men by the year, that the price of labor per day would average considerably less than it does at present in the summer time, which, in the region I have selected, is about one dollar and fifty cents per day for a first-rate hand.

One of the first merchants and manufacturers of France told me, that with wages at three and a half francs per day, the value of labor in a hundred kilogrammes of sugar should not exceed four to four and a half francs. That is, with wages at sixty-six cents per day, the cost of labor should be less than four mills per pound.

By the preceding tables the cost of labor at sixty-six cents per day varies in a pound of sugar from four to seven and one tenth mills in France. The average is not far from $5\frac{8}{10}$ mills per pound.

If the same amount of labor be required here as the average of France, and its value be three times greater, or two dollars per day, then the average cost of a pound of sugar from beets yielding seven per cent., will be five and one fourth, instead of four cents, per pound.

I herewith present a table showing the results that I have no doubt can be attained in Illinois by a company with $300,000 capital, of which $200,000 shall be appropriated for buildings and machinery, and $100,000 reserved for working capital.

EXPENSES.

24,000 tons of beets, . . .	@ $4.00 . . .	$96,000
Labor, 225 men, 150 days,	@ $1.75 per day,	50,625
Salaries, ...		10,000
Coal, 3,000 tons, ...	@ $1.50, . . .	4,500
Sacks for pulp, 8,000 . .	@ $1.00, . . .	8,000
Bone-black waste, ...		7,500
Insurance, ...		2,000
Lighting, ...		750
Lime, metals, barrels, rasp blades, repairs, &c.		15,125
		$194,500

RECEIPTS.

1,680 tons sugar (yield calculated at 7%), at $200 per ton, or $8\frac{9}{10}$ cents per pound, .	$336,000
720 tons molasses (yield calculated at 3%), at $10.00 per ton, or 4 cents per gallon,	7,200
4,800 tons of pulp, at $2.00 per ton (equivalent to hay @ $6.00 per ton), ...	9,600
	$352,800
Less expenses, ...	194,500
Profit equal to 52% on capital, ...	$158,300
From which is to be deducted for local taxes and internal revenue, ...	10,000
Net profit, being nearly 50% on capital, .	$148,300

It will be seen that the yield of sugar is placed at

seven per cent. I have no doubt it would be more, for by the method recommended, and which is in use in France, the yield is eight per cent. The price of sugar is also calculated at $8\frac{9}{10}$ cents per pound, but samples made by the process referred to are declared to be now worth an average of thirteen cents.

The value of the molasses I have placed at four cents per gallon, but it will produce twenty-five per cent. of its weight in 90° alcohol, and the market value of a material that will give that result is certainly not less than twenty-five cents per gallon.*

I have placed the market value of the pulp at two dollars per ton, at which price it has been ascertained, by years of experiment, to be equivalent to hay at six dollars per ton; therefore it cannot be said that the estimate is too high.

On the other hand, beets are charged at four dollars per ton, upon which there is little doubt a saving of fifty cents per ton, or twelve thousand dollars, could be effected. On pages 26 to 39 the probable cost of beets is discussed. There can be little doubt that the actual cost to the farmer will rarely exceed three dollars per ton, even with small crops, while with twenty or thirty tons per acre, the larger of which is by no means an uncommon yield, the cost would be from one dollar and a half to two dollars a ton. Manufacturers could

* The molasses contains from forty-five to fifty-five per cent. of crystallizable sugar. Until recently no economical method for its extraction was known. Last year, however, three or four establishments were erected in Europe for that purpose, and I have been assured that nearly all the sugar can be extracted at a cost of three and a half cents per pound.

certainly raise their own beets at three dollars per ton, and probably at considerably less.

In fact, there can be no doubt that the estimated expenses are placed sufficiently high, being at the rate of $4\frac{9}{10}$ cents per pound of sugar, or $1\frac{8}{10}$ cents higher than in the French manufactory, which it is proposed to copy; while excluding the item of labor, the balance of expenses would be less here than in France. The actual expenses for labor in the French manufactory are less than one half a cent per pound, and $1\frac{8}{10}$ cents per pound has been allowed as the excess of cost here over that in France.

I present below a table showing the estimated result, with the yield of sugar as great as in the French establishment, namely, eight per cent., provided it were sold at its present market value, say twelve and a half cents per pound, and the molasses at twenty-five dollars per ton, or ten cents per gallon, which is less than half its actual value for distillation.

1,920 tons of sugar at $12\frac{1}{2}$ cents per pound,	$537,600
720 " " molasses at $25 per ton, . .	18,000
4,800 " " pulp at $2 " " . .	9,600
	$565,200
Less expenses, ...	194,500
Profit (equal to 123 per cent on capital), .	$370,700
Or, deducting taxes and internal revenue, .	16,000
118 per cent, ...	$354,700

By the poorest methods prevailing in Europe six

per cent. of sugar is obtained. By the best processes nine per cent. of sugar and two and a half per cent of molasses can be and repeatedly have been extracted from beets containing twelve and a half per cent. of saccharine matter, which is the amount in the beets raised in Illinois on the first experiment. I submit, therefore, the accompanying table as an indication, on the one hand, of a result that is *possible to be realized*, and also, on the other, of a result that in the present state of the art is certain to be at least *equalled.*

In this table sugar is credited at ten cents a pound, molasses at ten cents per gallon, and pulp at two dollars per ton. Expenses are reckoned as in the preceding table on page 49.

TABLE

Showing the Products of Sugar from 24,000 *Tons of Beets, yielding Six, Seven, Eight, and Nine per cent., with the Amount and Percentage of Profit on a Capital of* $300,000. *Taxes and Internal Revenue not deducted.*

Yield per cent.	Yield of sugar. } tons.	Profit, dollars.	Profit per cent.
6	1,440	$152,660	50 8/10
7	1,680	206,420	68 8/10
8	1,920	260,180	86 7/10
9	2,160	313,940	104 6/10

On pages 40 to 48 I have discussed fully the probable cost of manufacturing beet-root sugar, and have arrived at the conclusion that under no circumstances, with a *yield of seven per cent.* of sugar, can the cost exceed $5\frac{1}{4}$ cents per pound. My belief is that it would be less, say $4\frac{3}{4}$ cents at the outside. But if it cost $5\frac{1}{4}$ cents, and sold at ten, there would still be a profit of ninety per cent.

After making all allowance for contingencies that I can imagine as possible to arise, I have not the slightest doubt that there can be realized on the manufacture a profit of at least eighty per cent. on the capital invested.

In a conversation with a French gentleman, a manufacturer of sugar machinery for all parts of the world, and who is also largely interested (and with most favorable results) in the manufacture not only of cane sugar in Martinique, but also of beet sugar in France, in Germany, in Poland, and in Russia, he gave it as his opinion, that the beet was destined to become the great sugar-producing vegetable of the world, for the reason that it can be cultivated in the temperate latitudes, in countries of dense population, and consequently in close proximity to the consumers of sugar. In his judgment sugar can be produced from it as cheaply in Europe or in the United States as it can be from cane in the West Indies or Brazil. And even if that position were not tenable, the expenses of transportation are so great as to render it absolutely certain that sugar produced from the cane cannot compete with beet sugars in the markets of Europe or the United States.

The "Journal des Fabricants de Sucre" says, that "the season of 1865–6 developed the success of two highly important processes, namely, the immediate carbonitation without defecation of the juice as it came from the presses, and the perfection of the operation of the improved filter presses. In the factories, where these new methods were employed, their superiority was marked in comparison with the old system, by which, late in the season, it was almost, and oftentimes quite, impossible to make good sugar. Beets that could not be successfully worked by the old process were brought to the new establishments, where sugar of beautiful quality, fit for direct consumption, was readily produced. And what was still more remarkable, in as great proportions upon the amount of beets worked as in the beginning of the season."

The entire success of these processes, which, seeing in operation, I have recommended the adoption of, has created the greatest excitement among the manufacturers in France. The opinion is there entertained that their employment will not only increase the average yield of sugar at least one per cent. on a hundred pounds, but also improve the quality of the sugar several numbers.

The remarkable results produced by these improvements have attracted the attention of Englishmen ; and the probability is, that the manufacture of beet sugar will yet be established in Great Britain, the country that not only tried to strangle the industry at its birth, but also, when it had been successfully established on her own soil, gave notice to the manufacturers, through its government, that an excise of five cents per pound would be placed upon their production, upon the ground that it would interfere with the prosperity of their West India possessions !

PRODUCTION OF SUGAR IN VARIOUS COUNTRIES.

RAMON DE LA SAGRA, in his work "Cuba en 1860," states that the average production of sugar per acre from the cane in that island was . . 1,709 lbs.

The highest, ...	7,980 "
" lowest, ...	1,257 "
Martinique average, ...	1,587 "
" highest, ...	1,900 "
Porto Rico average, ...	3,950 "
Reunion lowest, ...	1,100 "
" highest, ...	9,625 "
" average, ...	3,200 "
Mauritius, ...	8,562 "
Java, ...	4,166 "
I will add that the product in Louisiana before the war was about	1,100 "
In Germany, the average production from beets is about .	2,100 "
In France, average, ...	2,200 "
" " highest, ...	5,000 "

It will thus be seen that an acre of land produces from beets a larger average amount of sugar in France and Germany than is produced from cane in Cuba, Martinique, or Louisiana. In Mauritius the system of cultivation is good, but it is a matter of notoriety that the sugar of Mauritius cannot compete with beet sugar in France, notwithstanding it has an advantage over the latter in the French ports of five francs the hundred kilogrammes, or 4.3 mills per pound.

M. De la Sagra gives the following figures, showing the amount of sugar produced to a "hand" upon several of the best plantations in Cuba: —

La Ponina,	. 4,238 lbs.	Flor de Cuba,	6,430 lbs.
Conchita,	. 4,413 "	Delta, . .	7,062 "
St. Martin,	. 4,512 "	Las Canas, .	13,327 "

On some well-ordered estates, both in France and in Germany, the production of sugar to a "hand" exceeds 14,000 pounds.

The production of sugar at Martinique in 1832 was 30,000 tons. In 1850, in consequence of emancipation, it fell to 15,000 tons. In 1864, the production again reached 30,000 tons. Emancipation produced a similar result in Guadaloupe. In Reunion, by reason of immense importations of Coolie labor, production has increased fourfold since emancipation; but intelligent observers see that Coolie labor is but another form of slavery, for which reason the supply must cease. It does not, like slavery, reproduce laborers, for ninety to ninety-five per cent. of the Coolies are males. The increased production is also due to an extended area of cultivation, and not, as in Mauritius, to improved methods of culture. In fact, some of the most intelligent planters in several of the French colonies have abandoned sugar cane, and cultivate other crops.

ATTEMPTS TO MANUFACTURE BEET SUGAR IN THE UNITED STATES.

SEVERAL attempts on a very small scale have been made, within the last thirty years, to manufacture beet sugar in this country; but with one exception, so far as I can learn, they were made when the industry was in its infancy, and when prices were much lower than they are at present, or are now likely to be.

Those attempts were not crowned with commercial success; but the results produced were such as to demonstrate, beyond the shadow of a doubt, that beet sugar can now be made in this country with the most absolute certainty of success.

The attempt, of which there is now to be obtained the most complete published account,* was made at Northampton, in the valley of the Connecticut, in the years 1838–9, by David Lee Child, and the "Northampton Beet-sugar Company." The company were the successors of David Lee Child, to whom the Massachusetts Charitable Mechanic Association, at their second exhibition, in 1839, awarded a silver medal.

In their report the Association say, "The crude or raw sugar is well made, dry, and of good grain. The refined shows that this article can be made of as good quality as sugar from the cane."

On the 5th of December, 1839, the "Massachusetts

* The Culture of the Beet, and Manufacture of Beet Sugar, by David Lee Child, 1840.

Agricultural Society" awarded a premium of one hundred dollars to the "Northampton Beet-sugar Company, for beet sugar."

On the 13th of November, 1839, Hon. Levi Lincoln, president of the "Worcester County Agricultural Society," addressed a letter to Mr. Child, who had sent him a box of sugar for exhibition. The box arrived too late; but the following extract from Mr. Lincoln's letter indicates the quality of the sugar: "Availing of your kind permission, samples of the sugar were submitted to the inspection of several gentlemen. The *brown* sugar was found to be pure, very sweet, and entirely free from any bad taste, and its quality, in every respect, was highly satisfactory.

"The refined or lump sugar seemed not so well granulated as is desirable. Still we are well satisfied that, as an experiment in the manufacture, it is highly encouraging, and we all felt that the country was largely indebted to your intelligence and enterprise in demonstrating, beyond all question, how entirely this application of domestic industry is at her command."

In May, 1839, Mr. Child received a letter from Martial Duroy, of Boston, confectioner, from which the following is an extract: —

"Having, while in France, heard the confectioners in general deprecate the use of beet sugar in their work, I was naturally a little prejudiced against it when I was called upon by you to make some confectionery for the 'Ladies' Anti-slavery Fair.' I was pleased to find, upon trial, that your raw sugar was extremely easy to clarify, and that it grained freely. These attributes of good and pure sugar reconciled me at once with it, and I made a variety of confectionery as easily and as handsome as with the best Havana. But its power of crystallization is particularly interesting, as it is upon this that depends its successful transformation into loaf sugar; and as far as a pretty considerable experience goes to establish it, I think beet sugar obtained by your process does crystallize, both easily and abundantly, forming at will coarse or fine grains, peculiarly brilliant, and giving, by far, a smaller quantity of molasses in the process of refining than cane sugar of a corresponding quality. I found also the molasses of a pleasant taste, and well adapted in its chemical composition to culinary purposes."

Mr. Child says that the best result he obtained from one hundred pounds of beets was seven pounds of sugar and three and one third of molasses; that "the sugar was of excellent quality, free, even in its raw state, from any bad taste, and of a pure and sparkling white when refined. Old and extensive dealers have pronounced it in both states capable of successful competition with any sugars in the market."

The quantity made was about 1300 pounds.

Mr. Child satisfied himself, from the result of the labors of 1838–9, that "the raw sugar can be obtained without any bad taste, and fit for immediate consumption; that American beets, though generally inferior to the European in saccharine richness, can, by suitable culture, be made inferior to none."

He says, "The sugar grained in a few hours; drained well and is not inferior in flavor or appearance to the finest West Indies Muscovadoes. The quality of the molasses has been a matter of utter surprise to us. In France the molasses is considered of no value except for feeding to animals or for distilling, and it sells for four or five cents per gallon. The molasses from the sugar in question is of a bright amber color, and so pure and pleasant as to be preferred by many to any but sugar bakers'." He says, "It will be readily conceived that a small establishment, dependent upon farmers for material, paying for it twice the cost of its production, and executing by hand several heavy and tedious operations, which ought to be performed by steam, water, or horse power, cannot furnish accurate data for determining the expense of making beet sugar. The actual cost when the material was good has been eleven cents per pound, the pulp and manure not taken into account. We are of opinion that, with proper and sufficient means, beet sugar may be manufactured in the United States at four cents per pound. When the manufacture shall have become domesticated among us, it will probably be produced at a cost less than that."

In relation to the effect of a beet crop on succeeding crops, Mr. Child says, "In Northampton wheat has succeeded beets the present season with rather striking success. A farmer let a field abutting on Connecticut River on shares. On a part of it he raised beets last year, and on the other Indian corn. The whole was equally manured. The corn yielded seventy-five bushels to the acre, and the beets were tolerably weeded. The wheat was harvested, and his share delivered in the barn without any attention to it on his part. In due time a laborer was employed to thrash it. This person, after thrashing a quantity, observed to his employer that the wheat on one side of the loft thrashed easier, and had a better berry and brighter straw, than on the other. Upon examination it was found that the former had been produced upon the beet, and the latter upon the corn, section of the field, but with this difference, that the beet grew nearest to the river, where it is considered that wheat is most likely to blast. We had the advantage of examining these wheats, and the difference was clearly such as the thrasher had stated. The proprietor found a difference of three and a half pounds per bushel in the weight. We presume that the difference in the flour would be found much greater, because, the grains of the inferior wheat being smaller, it would require more of them to fill a measure; and as the shrunk grains have the same quantity of skin as the large, and as it is the skins which make bran, it follows that the superiority remarked would appear still more signally if the two samples were ground and bolted."

Mr. Child, in a note, remarks, "Mr. Harrison O. Apthorp, of Northampton, — one of the earliest cultivators of the sugar beet in this country, — has informed us of the remarkable growth of herdsgrass as a successor of sugar beets on his grounds. The crop was pronounced by the oldest farmers in Northampton village superior to any of the kind they had ever seen in the meadows."

Several years ago, beet sugar, of very fine quality, was made by the society of Shakers at Enfield, but upon too small a scale, and by too crude a method, to ascertain fairly the price at which it could be produced.

In 1863-4 the brothers Gennert, of New York, conceived the idea of manufacturing beet sugar. Mr. Thomas Gennert visited Europe for the purpose of studying the methods there employed. Upon his return, the firm selected the prairie lands in the town of Chatsworth, Livingston County, Illinois, purchased 2300 acres, erected buildings, and commenced the cultivation of beets. In process of time they gathered their crop, which, owing to the drought, and also to the unfavorable method of planting, yielded only ten or twelve tons to the acre. The beets were of excellent saccharine properties, containing twelve and a half per cent. in sugar. The heavy outlay required exhausted their means; or, to use their own words, "We started on too large a scale for our purse, which gave out too soon, before the machinery which was required for a successful working was finished; but experience has shown us sufficiently that sugar enough is contained in the beets, and that it can be got out. With our imperfect, or rather incomplete, machinery, we extracted seven per cent. in melado. Those beets would average, with complete machinery, nine per cent."

The Messrs. Gennert have put their property into a stock company, called the "Germania Sugar Company," and have six hundred acres of land in cultivation with beets this season.

I submit their estimate of the profits of working one hundred tons of beets per day, with the following productions of sugar, on a capital of $200,000: —

At 6 per cent.,	. .	73 per cent. profit.		
7 "	. .	91	"	"
8 "	. .	109	"	"
9 "	. .	127	"	"

General Advantages of Beet-sugar Manufacture.

The "Journal des Fabricants de Sucre," in its issue of December 8, 1864, says, "We find that the abolition of slavery in America and the West India Islands, which seems to us the inevitable result of the America war, at the same time that it increases the demand for sugar must diminish the supply about 500,000 tons. The production of Louisiana will be destroyed, that of Cuba diminished one half or one third, and that of Brazil will be reduced. How is this deficiency to be supplied? The consumption of the United States is nearly as large as that of Great Britain, and they will probably be driven by necessity to manufacture sugar from the beet, the processes for which they can learn of Europe. As for France, Belgium, and Germany, they can easily double or triple their production; for it does not require long preparation of the soil to produce beets. Capital is abundant for such an enterprise; and even at the present rate of increase, production doubles every ten years."

"England may fear that the manufacture of beet sugar in Great Britain would prejudice her colonial interests; but some of her statesmen foresee its introduction." The editor predicts that the effect of the change in the sources of supply would be to diminish, and not to enhance, the price of sugars. He goes on to say, "The North and the South may fight as long as they like. The 4,000,000 slaves in the Southern States may be freed, the 400,000 negroes in Cuba may also be emancipated, as well as those of Brazil. The African slave trade may stop, drought and insects may continue to ravage the sugar plantations of Reunion and Mauritius, but sugar will not become scarce in Europe for all that. We shall continue to be supplied by our own admirable industry, whose advantages and development we have set forth."

In a later issue the probability is discussed of the United States continuing to import annually 300,000 to 400,000 tons of sugar from Cuba and Brazil, "when they have the ability to supply all their wants with beet sugar from their own soil, not only with certainty of profit to the manufacturer, under the existing tariff, but also with advantage to the whole country, because of the unreliability of the cane crop of Louisiana, which never ripens, and which at any rate is certain to be paralyzed for the next ten years.

"But even if the duties on foreign sugars should be abolished, the advantage would be on the side of the beet-sugar manufacturer, who will probably have less need of protection than the Louisiana planter.

"The people of the Northern States will not long defer the cultivation of a plant which contains so much sugar that it will soon teach them to forget that which was formerly produced upon the banks of the Mississippi. As to the competition of Cuban and Brazilian sugars, they have no more cause to fear it than have the beet-sugar makers of France and Germany, where the economical conditions are far less favorable than those of the Northern and Western States."

The beet-sugar industry has been of vast benefit to Europe. Notwithstanding the high protective policy to which it owes its existence, and which, as a matter of course, was pursued for a time at the expense of the public, which paid higher for sugar than it would otherwise have done, yet there is no question that sugars have been cheaper throughout the world for the past fifteen years than they would have been had the industry not existed.

Formerly the production of sugar was a monopoly confined to the tropics, where its possession, combined with the cheapness of land and the system of slavery, fostered in planters and manufacturers an extravagant, shiftless, and costly method of manufacture.

The vast improvements that science has brought to bear on the chemistry and mechanics of beet-sugar production in Europe have awakened the planters and manufacturers of the tropics to the necessity for progress, if they desire to retain their supremacy.

Almost all the improvements made in cane-sugar manufacture in the last fifteen years owe their origin to the beet-sugar establishments of France and Germany.

The effects produced upon agriculture in Europe by the cultivation of beets for sugar and alcohol have been astounding, and the importance of the interest is now everywhere acknowledged.

In the cane-sugar countries upon the territory surrounding a sugar establishment no crop is to be seen but the cane, while cattle and sheep are few. In the sugar districts of Europe, on the contrary, the fields in the vicinity of a sugar manufactory are covered with the greatest diversity of crops, among which are beets, wheat, rye, oats, barley, corn, rape, flax, tobacco, and all the cultivated grasses. Every field is cultivated close up to the road-side, and the stables are filled with fine cattle, sheep, horses, and swine.

No farmer needs to be told which system is the best and most enduring.

M. Dureau, author of several valuable works on beet sugar, and also the editor of the "Journal des Fabricants de Sucre," says "The cultivation of the beet is getting to be highly popular.

"The president of an agricultural society is sure to gain all hearts when he talks about beets. No agricultural newspaper can abstain from entertaining its readers with accounts of the precious plant, and there is no farmer who does not introduce it into his fields with the view of its conversion either into sugar or alcohol. Everybody sings its praises; and surely none have a better right to join in the concert than we, who have always been its advocates for the sake of the industry with which it is allied."

A French writer, after having demonstrated the importance of the beet-sugar industry to agriculture, in urging its extension, says, "Who would believe that England, with her poor soil, her wet climate, and her pale sun, could produce crops of grain double ours, and that the yield of her fields surpassed that of the luxuriant plains of Lombardy? The perfection of her agriculture explains this wonderful production. So does the progress of the manufacture of beet sugar explain how the cultivator of the north can extract as much sugar from a hectare of his cold and wet land, as the indolent Creole from the rich soil of the Antilles, bathed in sweet odors and in sunshine."

The basis of the agriculture of England is the turnip. In the best cultivated districts of France, it is the beet. M. Barral, a celebrated writer on agriculture, says, "I did not find any good crops except in those countries where an industrial culture prevailed, which is especially the case in those where the beet is cultivated."

Another writer says, "Of all species of industry which it is desirable to see extended in France, the manufacture of sugar and alcohol occupies the first rank. Branches of industry which are pursued in the winter deserve to be supported, because they give employment to laborers who work in the fields in summer, and thereby enable them to increase the amount of their yearly wages."

Another writer says, that "all cultivators and economists are unanimous in recommending the cultivation of the sugar-producing plant, which is the source of deep tillage, heavy manuring, and increased production. No one believes now that it exhausts and impoverishes the soil, or that it hurts other crops: these are the prejudices of a by-gone age, which science and practice have banished, to set up in their place a recognition of benefits of the highest order produced by the culture of the beet."

M. Dureau says, "The manufacture of beet sugar was formerly charged with being a local industry. To-day it no longer deserves that reproach, for it is not alone in the north of France that it is pursued; but it has penetrated into the east, the west, and the south, — into Germany, Russia, Italy, Austria, Spain — everywhere."

Another says, that "everywhere the beet is cultivated in France, land advances in value, and the wages of workmen take the same direction."

"All Europe, though France has contributed the largest and most glorious part towards the accomplishment of the result, is destined to become a great sugar-producing country, not less important than those where they cultivate the cane, which many believed to be the only plant suitable for the production of sugar, that precious food, of which people of the present age are such large consumers. Why should not sugar, which the mysterious forces of nature have secreted in the beet, be extracted from it, and the soil, prepared for new harvests, and rendered doubly fertile by the thorough cultivation it demands, furnish increasing quantities of food for man, and for beast? It is the triumph of industry."

L'Echo Agricole says, that "all farmers who obtain first prizes at the agricultural exhibitions are either sugar manufacturers, distillers, or cultivators of the beet. Those who have adopted this branch of agriculture, either as proprietors or tenants, have really obtained astonishing results. They would be surprised if they did not carry off all the first prizes at the public exhibitions, and were consequently mentioned in the official reports of the government."

M. Vallerand, who took the first prize in the Department of Aisne, bought, in 1853, a farm of eight hundred and thirty-two acres, the sales of produce from which amounted to $8,000. In 1859 it produced $41,200. M. Dargent, who took the first prize in the Department of Seine Inferieure, cultivated only fifty acres. He so increased the production of this farm that he obtained 154,000 pounds, or 68 tons and 168 pounds, of beets from a single acre. His yield of wheat was 43½ bushels, and of oats 59½ bushels, to an acre.

M. Hary, Pas de Calais, obtained from two hundred and ninety-five acres 5,225 bushels of wheat, 2,500 tons of beets, and fattened 150 head of cattle.

The culture of the beet involves the necessity of deep ploughing, heavy manuring, and thorough weeding. The pulp from which the juice is extracted in the manufacture is an excellent food for cattle, the number of which has been increased, in the districts devoted to that industry, from eight to ten fold since the introduction of sugar making.

The cattle furnish an immense amount of manure, which, applied to the deeply-ploughed and well-weeded beet lands, enhances their productiveness for the cereal crops.

In 1853, when the emperor and empress came to Valenciennes, a triumphal arch was erected, with the following inscription: —

SUGAR MANUFACTURE.

Napoleon I. who created it.	*Napoleon III. who protected it.*
Before the manufacture of beet sugar, the arrondissement of Valenciennes, produced 695,750 bushels of wheat, and fattened 700 oxen.	Since the manufacture of beet sugar was introduced, the arrondissement of Valenciennes produces 1,157,750 bushels of wheat, and fattens 11,500 oxen.

The brothers Fievet have a model farm of 552 acres at Masny, which is considered the best in France. They are sugar manufacturers, and fatten 800 head of cattle and 3,000 sheep every year. I visited there last winter, and spent a day in their manufactories and on their farm. They attribute their success as cultivators to the immense amount of manure that the beet pulp enables them to make, to the improved condition of the soil, and also to the increased amount of profitable service of the land, consequent upon beet culture, no fallows being required.

They have cultivated the farm for thirteen years: the crops are beet, wheat, oats, rye, and hay. I shall give some of the results of the eleven years preceding 1864. The average amount of land in oats had been thirty acres. In 1853 the crop was 45½ bushels, in 1862 nearly 92¾ bushels, and the average for the whole time within a fraction of 70 bushels to the acre.

The crop of straw increased in like proportion, and averaged two tons to an acre. In 1863 it was nearly three tons.

The crops of rye improved in a still greater ratio — increasing from 17 to 34½ bushels per acre, averaging nearly 30 bushels, with two tons of straw to the acre.

The average crops on 156 acres of wheat had been over 36½ bushels to the acre.

Parts of the land had sometimes produced 67⅝ bushels to the acre, and no portion had ever yielded less than 20½ bushels. The yield of hay had been over three tons; and of beets twenty tons to an acre.

In 1865, thirty, thirty-five, and even forty tons of beets were raised on an acre.

As to the cost of producing these crops, the Messrs. Fievet stated that the thorough cultivation of the ground for beets reduced the cost of cultivating succeeding crops enormously.*

Thus, after deducting the proceeds of the straw, their oats cost them less than thirty cents, their wheat less than sixty cents, and their rye less than thirty-eight cents, per bushel.

This they attribute to underdraining, to the use on the beet crop of lime, either pure or the carbonate of lime from the filter presses of the factory, to the liberal application of other manures, to deep ploughing, thorough weeding, and cultivation. The grain crops are not manured, and the ground is so thoroughly prepared

* The subjoined table shows approximately the average yield of certain crops per acre in twenty-three of the United States, in the year 1865, according to the Report of the Department of Agriculture for January, 1866 : —

Crops.		Highest average yield.		Lowest average yield.	
Wheat, ...	13½ bush.	Minnesota,	20⅗	Kentucky,	7¼
Rye, ...	15 "	Kansas,	23	Delaware,	7
Barley, ...	23¾ "	Vermont,	28¾	Mass.	19½
Oats, ...	31¾ "	Minnesota,	41½	Delaware,	12
Corn, ...	36½ "	Nebraska,	46½	Delaware,	16½
Buckwheat, . . .	19¼ "	Nebraska,	26⅔	Delaware,	10½
Potatoes, ...	113 "	Minnesota,	197	Kentucky,	59½
Tobacco, 16 States, .	906 lbs.	Conn.	1,350	Kansas,	533
Hay, ...	1½ tons.	Nebraska,	2	Maine,	1
Sorghum molasses, 18 States,	110¾ gals.	Kansas,	129	New York,	75

The productions of the farm at Masny vastly exceed those of the States named. The explanation is to be found not in the soil or the climate, but solely in the cultivation.

by the beet for succeeding crops, that a single light ploughing suffices for the grain, which is all sowed in drills by a machine.

Before the introduction of the sugar industry into France, workmen in the country, by reason of a lack of employment, were so constantly emigrating to the city, that government instituted inquiries to ascertain the cause, and also the best method of preventing it. Now, the natural tendency of workmen to seek the capital is not noticed in the sugar-producing districts, where the industry gives ample and well-paid employment to all, both in summer and in winter, and where crime and pauperism have sensibly diminished.

Agriculture was looked upon as the calling of peasants, requiring little intelligence and no education. It is far otherwise now, and to be successful as a farmer involves the necessity of having a good education. The introduction of sugar-making into France, and the intimate relation between that industry and agriculture, called for improved methods of culture, and a more intelligent and scientific application of labor. Intelligence and education were decentralized for the benefit of the whole country; capital also lent its powerful aid, and agriculture made rapid progress, while the condition of the laborers also was materially improved.

Louis Napoleon, the present emperor of the French, when he was imprisoned at Ham, in 1842, said of the beet-sugar industry, in his "Analyse de la Question des Sucres," "It retains workmen in the country, and gives them employment in the dullest months of the year; it diffuses among the agricultural classes good methods of culture, calling to their aid industrial science and the arts of practical chemistry and mechanics. It multiplies the centres of labor. It promotes, in consequence, those sound principles upon which rest the organization of society and the security of governments; for the prosperity of a people is the basis of public order. * * *

"Wherever the beet is cultivated, the value of land is enhanced, the wages of the workmen are increased, and the general prosperity is promoted."

In another place the same author puts the following words in the mouth of the sugar industry: "Respect me, for I improve the soil. I make land fertile, which, without me, would be uncultivated. I give employment to laborers, who otherwise would be idle. I solve one of the greatest problems of modern society. I organize and elevate labor."

The conclusions to which I have arrived are, —

That the skill, which is the result of the experience of more than a century, and which has made France independent of foreign countries for her supply of sugar, is available for us to-day.

That the manufacture of beet sugar can be successfully transplanted from France to the United States.

That sugar can be produced in this country from the beet nearly if not quite as cheaply as it can be from the cane in Cuba, or any other country.

That the protection of transportation alone is sufficient to render it impossible for the sugar of tropical climates to compete with beet sugar in the United States.

That as the climate of the Southern States does not permit the cane to ripen, and as the yield of sugar from unripe cane is comparatively small, it is impossible to make sugar from cane in the United States so cheaply as it can be made from beets.

And that at present prices beet sugar can be manufactured in this country at a profit of from eighty to one hundred per cent.

By the new internal revenue law beet sugar enjoys a protection over the sugar of the cane of from one to two cents per pound in currency.

Duties on foreign sugars are from three to four and a half cents per pound in gold.

The necessities of government, and the very apparent advantages arising from introducing the manufacture of beet sugar into this country, render it probable that the protection now accorded will be maintained for the present.

The cost of transportation from the seaboard to Illinois is an additional protection on sugar raised in Illinois of about one cent per pound.

The amount of beets raised in France in 1865 could not have been, on 297,000 acres of land, less than 5,000,000 tons, producing at least 1,000,000 tons of pulp — an amount sufficient to feed 90,000 cattle or nearly 1,000,000 sheep for one year, or to fatten in the winter months nearly three times that number. It also furnished agriculture with more than 1,500,000 tons of manure. In an agricultural point of view, the effect produced by the culture of so much land in beets, and the application of the manure of so many cattle, with the consequent increase in the amount and value of subsequent crops, is perfectly apparent. The quality of wheat raised after beets is better than that usually produced; the ears are larger and heavier, the straw stronger, and not so liable to lodge. The berry is larger and brighter; its specific gravity is also greater, weighing from two to three pounds per bushel more than ordinary wheat.

But these effects are not all, even of those having an agricultural bearing, which the great industry produces. They are not confined to the comparatively narrow circle that surrounds the factory, in which are expended for beets and for labor large sums that foster industry, and scatter plenty in the surrounding villages. The distribution of these large amounts for labor and for the crop opens a better market for the productions of other branches of industry, agricultural, mechanical, manufacturing, mining, and commercial.

To till the land and to consume the pulp, many horses, as well as vast numbers of cattle and sheep, are required. These are purchased from other sections, for the departments in which the beet is cultivated are not *grazing* districts in which cattle are *raised*, but they are preëminently distinguished for *supporting* and *fattening* cattle.

The improved condition of the 70,000 laborers engaged in this industry, one fifth of whom are women and children, makes them larger consumers of tea, coffee, meat, clothing, — of all the necessaries of life. Their enlarged means place within their reach many hitherto unattainable luxuries.

The industry also calls into existence many establishments for the manufacture of agricultural tools. It gives employment to chemists and engineers; to machinists, founders, carpenters, blacksmiths, coppersmiths, wheelwrights, and plumbers; to woollen and linen manufacturers for the sacks it requires. It is a large consumer of coal, of iron, and of other metals, products of the mine. It contributes largely to the support of railroads and canals. It adds its quota to the extension of commerce. Finally, it pays to government an excise tax on sugar and alcohol of more than $27,000,000 per annum, without taking into account other taxes, state and local, that are assessed on the $45,000,000 that it has invested in buildings and machinery.

It has not only added immensely to the extent of arable land, but has largely increased the productiveness and value of that already cultivated. It has enabled France to produce more corn at less cost than she ever did before, and kept down the prices of all grains, of beef, and of mutton. At the same time it produces for man sugar, meat, bread, alcohol, potash, and soda; it furnishes nutritious food for cattle, sheep, and swine, together with hay and grain for the horse. In the opinion of eminent French statesmen, it has twice, within fifteen years, saved France from a famine.

The historian Thiers has called it "the Providence of the empire."

Effect of its Introduction into the United States.

The effect of its introduction into the United States would be to produce results correspondingly greater than have attended it in Europe, for here the consumption of sugar per capita is nearly four times greater, and the value of lands is not a quarter of those in continental Europe, while they are by nature far richer and more easily cultivated. The supply of coal is unlimited. The vast distances over which many farmers are obliged to transport their produce render it oftentimes impossible to dispose of their more bulky crops at a profit. The introduction of sugar-making would give them another and most profitable crop, for which they would have a home market. It would enlarge the local demand for other farm produce by interspersing a manufacturing with an agricultural population, to the great advantage of both. It would go far to change the present wasteful and necessarily unenduring system of agriculture, and to substitute for it another, founded upon more correct principles — a system self-sustaining and improving, rather than suicidal and degenerating.

The gold value of sugars imported into this country is nearly $80,000,000 per annum.

The annual consumption of sugar in the United States before the war was over 450,000 tons.

There is no doubt that within twenty years it will be more than 1,000,000 tons, for with the customary increase of population and the consumption per head that existed before the war, that amount would be required.

With a proper rotation of crops the production of that amount of sugar involves the cultivation of 4,000,000 acres of land, of which 1,000,000 would be in beets, the base of the system. It would give employment the year round, in the fields and in the mills, to more than 125,000 men, women, and children. It would require $100,000,000 to be expended in buildings and machinery. It would disburse annually $100,000,000 for labor and materials. It would require each year more than 1,500,000 tons of coal. It would fatten every year 400,000 head of cattle, or 4,000,000 sheep.

There is hardly an interest that it would injure, while it would be difficult to find one that would confer so many, so great, and so general advantages upon the country. It is destined to become one of the most important branches of national industry.

PART II.

THE BEET AND ITS CULTIVATION.

The beet is a half-hardy, biennial plant; its roots attain their full size the first year, but will not survive our winters in the open ground. Seeds are produced from transplanted roots, after which the plant dies.

Analysis of the Beet, according to Professor Payen.

	Per cent.
Water, ...	83.5
Sugar in solution, ...	10.5
Cellulose and pectose, ...	.8
Albumen, caseine, and nitrogenous matters, . .	1.5
Malic acid; pectine; gummy substances; fatty, aromatic, and coloring matters; essential oil; chlorophylle; asparamide; oxalate and phosphate of lime; phosphate of magnesia; silicate, nitrate, sulphate, and oxalate of potash; oxalate of soda; chloride of sodium and potassium; pectate of lime, potash, and soda; sulphur, silica, and oxide of iron, ...	3.7
	100.

Varieties of the Beet.

There are many different varieties, of which I shall describe a few of those employed for manufacturing and agricultural purposes.

LONG RED MANGEL-WURZEL.

Red Mangel-wurzel. — Marbled Field Beet. — ***Lawson.***

Burr describes this beet as follows: "Root fusiform, contracted at the crown, which in the genuine variety rises six or eight inches above the surface of the ground. Size large, when grown in good soil; often measuring eighteen inches in length and six or seven inches in diameter. Skin below ground purplish-rose; brownish-red where exposed to the air and light. Leaves green; the stems and nerves washed or stained with rose-red. Flesh white, zoned and clouded with different shades of red.

"The long red mangel-wurzel is hardy; keeps well; grows rapidly; is very productive, and in this country is more generally cultivated for agricultural purposes than any other variety. According to Lawson, the marbled or mixed color of its flesh seems particularly liable to vary: in some specimens it is almost of a uniform red, while in others the red is scarcely, and often not at all, perceptible. These variations of color are, however, of no importance as respects the quality of the roots. The yield varies with the quality of the soil and the state of cultivation, thirty and thirty-five tons being frequently harvested from an acre."

GERMAN RED MANGEL-WURZEL.

Disette d'Allemagne. — *Vilmorin.*

Burr describes it as " an improved variety of the long red mangel-wurzel, almost regularly cylindrical, and terminating at the lower extremity in an obtuse cone. It grows much out of ground; the neck or crown is comparatively small; it is rarely forked or deformed by small side roots, and is generally much neater and more regular than the long red. Size very large; well-developed specimens measuring from eighteen to twenty inches in length, and seven or eight inches in diameter. Flesh white, with red zones or rings. Leaves erect, green; the stems and nerves washed or stained with rose-red.

" For agricultural purposes this variety is superior to the long red, as it is larger, more productive, and more easily harvested."

LONG WHITE GREEN-TOP MANGEL-WURZEL.

Green-top White Sugar. — Long White Mangel-wurzel. — Disette blanche à collet vert. — *Vilmorin.*

" An improved variety of the white sugar beet. Root produced much above ground, and of very large size; if well grown, measuring nearly six inches in diameter, and eighteen inches in depth — the diameter often retained for nearly two thirds the length. Skin green, where exposed to light and air; below ground, white. Flesh white. Leaves green, rather large, and not so numerous as those of the white sugar.

"Very productive, and superior to the long red for agricultural purposes; the quality being equally good, and the yield much greater." (*Burr.*)

Vilmorin describes it as follows: "It is one of the best for sugar manufacturers. It has a smooth skin, grows beneath the surface, is rather large, and keeps well. Production, sixteen to twenty tons to an acre. It has been neglected lately in France, because there are so many kinds resembling it, which grow out of the soil, and are less profitable to sugar manufacturers. It is, nevertheless, superior to the collet rose."

LONG WHITE RED-TOP MANGEL-WURZEL.

Disette blanche à collet rose. — *Vilmorin.*

Vilmorin describes this beet as follows: "The betterave blanche à collet rose was formerly more extensively cultivated than at present, farmers having substituted for it the 'collet vert;' but the appearance of so many degenerate kinds of the latter has lately induced many farmers to resume the cultivation of the former. Its root is well shaped, smooth, long, and grows but little above the soil. Its flesh is white, zoned with red. It contained in 1860 about seven per cent. of sugar. In spite of this low percentage its cultivation in the north of France is increasing. By improved culture it produces larger quantities of sugar, approaching in richness to the standard of the 'collet vert;' it keeps well, and its color enables manufacturers and cultivators to recognize it readily."

YELLOW CASTELNAUDARY.

Burr describes it as follows: "Root produced within the earth, broadest at the crown, where its diameter is nearly three inches, and tapering gradually to a point, the length being about eight inches. Skin orange-yellow. Flesh clear yellow, with paler zones or rings. Leaves spreading, those on the outside being on stems about four inches in length; the inner ones are shorter, numerous, of a dark-green color, and rather waved on the edges: the leaf-stems are green rather than yellow.

"An excellent table beet, being tender, yet firm, and very sweet when boiled, although its color is not so agreeable to the eye."

Sarrazin describes it (betterave jaune) as "growing entirely beneath the surface, and having the form of a pear, not very heavy, but quite sugary, producing little foliage, succeeding well in poor soils, and yielding well where other kinds produce small crops. The stalks of the leaves have the same yellow color as the root."

YELLOW GLOBE MANGEL-WURZEL.

Betterave jaune globe. — *Vilmorin.*

"This is a globular-formed beet, measuring about ten inches in diameter, and weighing ten or twelve pounds; about one half of the root growing above ground. Skin yellow where it is covered by the soil, and yellowish-brown above the surface where exposed to light and air. Flesh white, zoned or marked with yellow, close-grained, and sugary. Leaves not large or numerous, rather erect, green, the stems and ribs paler, and sometimes yellowish.

"The yellow globe is one of the most productive of all the varieties, and though not adapted to table use, is particularly excellent for stock of all descriptions, as the roots are not only remarkably sugary, but contain a considerable portion of albumen. It retains its soundness and freshness till the season has far advanced, does not sprout so early in spring as many others, and is especially adapted for cultivation in hard, shallow soil.

"The yield varies from thirty to forty tons per acre, according to soil, season, and culture; although crops are recorded of fifty tons and upwards.

"On account of its globular form the crop can be harvested with great facility by the use of a common plough." (*Burr.*)

MAGDEBURG.

"The Magdeburg beet unites most of the qualities of the German race; its root is tapering, of middling size, with few accessory or lateral roots, and grows entirely beneath the surface, is white, and has a green neck. Its average yield is twelve to fourteen tons per acre in land where the white French sugar beet produces sixteen to eighteen tons.

"Experiments have shown it to be rich in sugar." (*Vilmorin.*)

IMPROVED VILMORIN.

"This kind, which is still in its infancy, is the richest of all, experiments having proved that it contains from sixteen to seventeen per cent. of saccharine matter.

"The neck of this beet is very large; the roots are generally irregular, of bad shape, and have many accessory roots; harvesting is difficult, especially in wet weather." (*Vilmorin.*)

IMPERIAL.

"The imperial beet is a native of Germany. It is said to contain thirteen and one half per cent of sugar. The root, which is carrot-shaped, has a green neck, is very long, and grows entirely beneath the surface." (*Vilmorin.*)

WHITE SUGAR.

White Silesian. — Betterave blanche. — *Vilmorin.*

"Root fusiform, sixteen inches in length, six or seven inches in its greatest diameter, contracted towards the crown, thickest just below the surface of the soil, but nearly retaining its size for half the depth, and thence tapering regularly to a point. Skin white, washed with green or rose-red at the crown. Flesh white, crisp, and very sugary. Leaves green; the leaf-stems clear green, or green stained with light red, according to the variety.

"The white sugar beet is quite extensively grown in this country, and is employed almost exclusively as feed for stock, although the young roots are sweet, tender, and well flavored, and in all respects superior for the table to many garden varieties. In France it is largely cultivated for the manufacture of sugar, and for distillation.

"Of the two sub-varieties, some cultivators prefer the green-top; others, the rose-colored or red-top.

The latter is the larger, more productive, and the better keeper; but the former is the more sugary. It is, however, very difficult to preserve the varieties in a pure state, much of the seed usually sown containing, in some degree, a mixture of both.

"It is cultivated, in all respects, as the long red mangel-wurzel, and the yield, per acre, varies from twenty to thirty tons." (*Burr.*)

Mauny de Mornay says, "The white Silesian beet is generally considered the best for the sugar manufacturer: it grows beneath the surface, has a small green neck, the stalks of the leaves are greenish-white; it yields less juice, but of a richer quality, than most other kinds; it contains salts in smaller proportions, keeps well, and resists frost better than others. It has the preference over all others with the manufacturers of sugar."

Characteristics of Beets for Sugar-making.

For the use of sugar manufacturers the kind of beet that can be cultivated with most advantage is that which is richest in sugar, and contains the smallest amount of alkaline salts. It is distinguished by the following characteristics: —

First. Its root must have neither the form of a carrot, nor of a tuber, but be shaped more like a Bartlett pear. It must be long and slender, gradually tapering, and free from large lateral roots.

Second. It must not grow above the surface of the soil.

Third. It must have a smooth white surface, and the flesh be white and hard.

Fourth. Its size must not be too large, and its weight not exceeding five to eight pounds.

The white Silesian beet, which is the one in general cultivation for manufacturers, unites most of these qualities; and of other kinds those are most preferred whose foliage is not upright, but broad-spreading and lying upon the surface of the ground. The roots of beets possessing this peculiarity grow entirely beneath the surface.

The beet, as a sugar-producing plant, is for the temperate latitudes what the cane is for the tropics; but besides its saccharine properties, it possesses others which render it even a greater acquisition to the human race than the cane.

It flourishes in almost any good soil; few plants are more hardy and tenacious of life, or have a wider range of cultivation.

It succeeds well in every country of Europe, from Italy to Norway, and from Spain to Russia.

In the United States it has been successfully cultivated in most of the states from Missouri to Maine, and would doubtless thrive in all. It is, however, a remarkable fact, that while the cane increases in saccharine richness as it approaches the equator, the reverse is the case with the beet, which up to a certain degree, north or south, secretes more sugar as it approaches the poles.

The northern limit of the successful culture of sugar beet on this continent is probably to be found at about latitude 50° to 52°, which is in Canada. In Europe it is successfully cultivated as far north as 60°.

Choice of Soil.

Although most countries and climates permit its culture, there is of course a choice of soil, if the highest development of saccharine qualities is desired.

The root of the beet penetrates deeply into the ground, and is abundantly supplied with fine fibres, through which it derives its nourishment.

The beet dislikes a too clayey, tenacious soil. Rocky or stony land must also be avoided, as it produces forked and misshapen roots, difficult to cleanse and rasp.

Soil charged with mineral salts is not suitable; for sugar beets easily absorb its saline and alkaline elements, which are obstacles to the extraction of sugar.

Marshy, swampy lands, and those in proximity to the sea, are unfavorable for the beet.

Wet lands are disadvantageous; but by a proper system of drainage, cultivation, and manuring, may be rendered suitable.

The beet flourishes best in deep, rich, loose, permeable soils, suitable for grains.

Light, rich, sandy ground furnishes beets dense, easy of preservation, and rich in sugar.

Calcareous soils are good, and the argillo-calcareous are better still.

Ground that is mellow, warm, and fertile, free from saline and alkaline constituents, not sour, and of a nature little liable to suffer from drought, easy to work late in autumn and early in spring, with a comparatively permeable subsoil, penetrable by the tap-root of the beet, that affords natural drainage, so that it may be worked soon after rains, is suitable for the crop in question.

The best colors for the soil are black and brown, provided the color is from vegetable mould, and not due to metallic elements.

A black soil warms more readily, and retains heat better, than that of any other color. This is favorable to the early development of the beet in the spring, which is important, as it tends to put the plant beyond the reach of summer drought, its long root penetrating deep enough into the earth to obtain the necessary supply of moisture. The "black soil" of Russia, which corresponds with much of our western land, is said by Professor Witt, of Munich, to be acknowledged the best in Europe for the sugar beet.

Count Chaptal, a great cultivator, as well as sugar manufacturer, says, "All grain-fields are more or less suitable for beets, but especially those having a depth of twelve or fifteen inches of rich vegetable mould. Fine, sandy, alluvial bottom lands, overflowed in winter or early spring, are favorable for the beet, and they need no artificial manure, as they are enriched by the inundations. Beets require to be planted on thoroughly cultivated land in which the sods are entirely rotted."

He was often compelled to sow a crop of oats on land newly broken up before he planted the beet, of which afterwards he often got two excellent successive crops.

When the soil was very light and deep he sometimes succeeded in getting a good crop on pasture land broken up in the fall, and planted with beets six months later; but lands in English grass, which were ploughed, and planted with beets, never produced a good crop the first year. It was always better the second year.

By intelligent, scientific, and well-directed labor almost any soil can be made suitable for the beet; but it is to be considered whether, in an economical point of view, it is judicious to force the culture of a plant upon a soil naturally unsuited to it. Heavy expenses will diminish if they do not entirely absorb the profits, even on large crops. This consideration is especially entitled to weight in the cultivation of sugar beets, for which it is best to select what is called in Europe a "natural beet ground."

A clayey, sandy subsoil, which retains moisture and the liquid manures, or a subsoil of marl, is favorable. On the other hand, a subsoil of gravel is unfavorable; so also is a subsoil of sand, unless the deposit of loam above it is at least two feet deep.

Drought in the season of early vegetation is pernicious, but after the plant gets well established it will bear extended dry weather. Too much rain, later in the season, increases the weight of the crop at the expense of sugar; or rather it diminishes the percentage of saccharine matter, the same amount existing, but in a less concentrated form. The beets are more watery, and consequently of less value for making sugar. Too much rain early in the season, when the plant needs warmth, is disadvantageous, and retards its growth.

M. Michael, in the "Journal de Chimie pratique," says, —

"1. The formation of cane sugar in beets is only favored by the proper concurrence of warmth and rain.

"2. Continued drought induces acid juices.

"3. The juices, during the period of storage in the silos (pits), are converted into grape sugar, which is uncrystallizable.

"4. Beets in the year 1859 (a very dry season) heated in the silos, and rotted sooner than those of the preceding year. This was the result of the drought and consequent formation of acid juices."

J. J. Fühling, a great Prussian cultivator, says, —

"My observations and inquiries satisfy me that in a climate warm and moist in summer, most lands are adapted to the beet ; that in a climate where the summers are very hot and dry, a strong and retentive soil is required ; and where they are colder and more humid, fields light and permeable produce better results for the cultivator.

"After planting is done, warm and moist weather in May and June favor the early development of the plant, which gives earnest of a good crop.

"With July and early August dry and warm, the production of good seed is probable.

"Continued and abundant rains in July and August insure a heavy crop. September dry produces beets rich in sugar ; but September wet makes them watery, and comparatively poor in saccharine matter, — not because the beets secrete their sugar in that month, but because with dry weather the beet ripens and its leaves begin to wither, while with continued rain the plant is stimulated to produce a second crop of leaves at the expense of the sugar contained in the root.

"The three principal periods of vegetation in the growth of the beet are marked by the successive formation of the leaves, the root, and the seeds.

"The first of these periods extends to July, during which time the leaves are rapidly developed, while there is but little increase in the size of the roots. The beet then remains in a state of comparative repose."

From the middle of July to the latter part of August the root increases rapidly in size.

Seed ripens in August.

From August to the middle of September, and sometimes until the 15th of October, the beet still grows, but increases more rapidly in weight than in size.

METHOD OF CULTIVATING THE SUGAR BEET * FOR THE MANUFACTURE OF SUGAR.

Having selected a suitable piece of ground that is already in cultivation, it should be thoroughly manured in the fall, the manure ploughed in to a depth of six or seven inches, and completely covered, taking particular care that the land is dry, for working wet land always develops in it gummy and sticky properties that subsequently interfere with easy cultivation. This superficial ploughing should be followed by a second, as deep as possible. A double Michigan plough would probably perform the work with a single oper-

* The instructions here given are exclusively for the cultivation of beets destined to be manufactured into sugar. The cultivation of forage beets for feeding stock is quite different, particularly in relation to the distances at which the plants should stand apart.

ation. The depth of the furrow should not be less than twelve inches, and if deeper, so much the better; for the root of the sugar beet requires a deep, rich bed, in which it can develop itself entirely beneath the surface of the soil. The part that grows above the ground contains no sugar, and if it rises much, is always cut off at the time of harvest, that course being mutually agreed upon in Europe by the manufacturer and cultivator.*

If the soil is ploughed to a sufficient depth, the root of the beet will not rise above the ground. The farmer consequently not only gets a larger crop, and of better quality, but the whole of it is marketable. Whereas if the ground is not properly ploughed, the beets rise, the part above ground is cut off, and is only used for feeding stock.

Deep ploughing therefore is of the greatest importance, not only for the beet, but also, as every farmer knows, for succeeding crops. It renders the soil mel-

* The portion of the root that grows out of the ground contains little or no sugar, but is rich in salts; therefore there is not only no good derived by the manufacturer from this exposed part, but a positive evil; for besides lessening the percentage of sugar contained in the whole root, the presence of the salts in the neck lessens still farther the percentage that can be extracted. This is so well understood, that in Germany, where women and children can be hired at ten or fifteen cents a day, they are employed in the factories to cut off from the beets, before they are rasped, every part of the crown and neck that grew above the surface of the soil. The portion thus cut off is fed to cattle. In France, where labor is higher, this custom does not prevail; but if the beets grow much above the surface, the necks are cut off at the time of harvest.

lower by more thoroughly exposing it to the action of the frosts. This is considered so important in Europe that the plough is often followed by laborers, who, with a spade, take out the earth from the bed of the furrow and lay it on the slice. Our subsoil plough would do that work cheaper and better.

If, after the fall ploughings, any weeds make their appearance before winter sets in, it is a good plan to pass over the field twice with a harrow, running the second time across the track of the first harrowing.

In Europe farmers use what they call an "extirpateur," which is an instrument with teeth sharp and strong, and about fourteen inches long, shaped like those of a cultivator. It differs from our "extirpator." It is used upon the stubble in the autumn, immediately after the grains are harvested, to extirpate the weeds, and is a very serviceable instrument. They usually pass twice over the fields with it, making the second track across the first. It is mounted on wheels, is of various sizes, and is drawn by two, three, or four horses.

If the land is so mellow as not to require a deep ploughing in the fall, manure is put upon the field, and the extirpator is passed two or three times over it. The land is then thoroughly cross-harrowed, and left until spring, when the treatment is the same as if it had been subjected to deep ploughing.

As soon as the ground is sufficiently warm and dry in the spring, it should be ploughed again, across the furrows of the preceding fall, to a depth of about eight inches, and again thoroughly cross-harrowed. If the nature of the land is wet, or if the upper soil is thin, it is sometimes thrown into ridges or beds in Europe. This method, the culture "en billon," finds many advocates even among those possessing lands of the deepest and most suitable soil. In case this method is adopted, only half the manure allotted to the field is used in the fall, and the rest is applied in the spring. The following is the method adopted: —

The portion of manure that is to be used in the fall is spread upon the land, ploughed in, and the field left in furrow through the winter. In the spring the field is worked up with the plough into ridges or "billons," between eighteen and twenty inches apart.

The remainder of the manure is applied, taking care to have it placed well at the bottom of the furrows. The ridges are then split with a plough, the manure in the furrows covered, and new ridges formed, which are then levelled with a light roller, and the seed sown in the usual manner, in the centre, directly over the manure.

Beets cultivated in this way are more apt to be forked than those raised by other methods.

The yellow globe ("jaune globe") is for that reason generally used in this culture, as its habit is to produce smooth and well-shaped roots. The advocates of this mode of culture claim that it produces larger crops, and is safer from the effects of drought than any other; but in my judgment the method is of doubtful expediency.

If the culture "en billon" is not adopted, then, after the spring ploughing and harrowing, the field is again gone over with the harrow turned upside down. This treatment serves better than rolling to smooth and pulverize the soil, and leaves it in admirable condition for the succeeding operations of sowing. All stones and clods that would interfere with the successful working of the seed-sower should be removed.

SOWING THE SEED.

The proper time to commence sowing is in the latter part of April, or as soon as the ground is in a fit state, being warm and dry, but at the same time sufficiently humid to promote rapid germination, and not so wet as to induce crustation or baking of the surface. Some European cultivators say that it should be done when the moon is on the increase.

Sowing in Europe is done both by hand and by machines; but as the price of labor in this country forbids the use of the former method, I shall give no description of it, although it is done by women and children very rapidly, and certainly possesses many advantages in countries where labor is low. I shall not describe either the method of transplanting the beet which prevails extensively in Germany, for labor is too high here to warrant the practice. Nor shall I give any description of the seed-sowers in use in Europe, because we have better ones in this country. I shall assume that machines will be used that sow several ranges or rows at a time.

The irregularity in size and shape of beet seed renders it necessary to subject it to certain treatment in order to facilitate the operation of sowing, and to prevent the clogging of the machine, the result of which would be to leave long spaces in the lines without any seed. This preliminary treatment also facilitates its germination, and in a measure guards it against destruction by insects.

The seed should be passed through a screen with meshes sufficiently fine to retain all that would not pass easily through the gauge that regulates the passage of seed in the machine.

The seed which do not pass must be rubbed between two boards, and partially crushed, in order to reduce those which are large and irregularly formed to a size that permits their easy transmission through the screen. After all the seed are by rubbing rendered sufficiently small to offer no obstruction to their easy sowing, they are steeped in the following solution: —

Dissolve nine ounces of sulphate of potash and an equal quantity of sulphate of lime in from four to five quarts of warm water. After this add five or six gallons of cold water. Of this solution use a sufficient quantity to cover the seed.

After having steeped for five or six hours, the liquid is drained off, and the seed are dried by putting them into a vessel either with wood ashes, slaked lime, ground plaster of Paris, or thoroughly pulverized guano, and mixing them together, so that each seed may be in a degree coated with the material employed. They are then spread until sufficiently dry to work readily in the machine. The machine should be set so as to sow the seed from one and a half to two inches deep, and in lines sixteen to eighteen inches apart, although some farmers make their rows fourteen and others twenty inches apart.*

* Beets planted a foot apart will produce about four tons more

The amount of seed required for an acre varies, of course, in accordance with the number of rows and the perfection of the seed-sower. It is certainly best to sow enough, for in seed-sowing apparent prodigality is often the truest economy, it being less costly to pull out superabundant plants than to sow a second time.

The farmer should bear in mind that the plants must finally stand from twelve to fourteen inches apart in the row. Knowing this, and the capacity of his machine, he can arrive at a pretty correct estimate of the amount of seed required.

In France the farmers employ from nine to thirteen pounds on an acre.

Too much pains cannot be taken to have the lines perfectly straight, and each passage of the machine over the field exactly parallel to the preceding one. "Marking," before the passage of the seed-sower, should be done with the very greatest care and exactness. This is of the utmost importance in every subsequent stage of cultivation, and cannot be too strongly urged. *For economical cultivation it is indispensable.* This is attained in Europe, and the lines are perfect miracles of straightness.

A strip of land sufficiently wide for the various machines and their teams to turn on should be left at each end of the field. In this country, where land is cheap,

per acre than if planted at a distance of eighteen inches; they will also be from half to one per cent. richer in sugar. But the lesser distance is not so well adapted to cheap culture, and the usual method is to have the rows sixteen to eighteen inches apart, and the plants twelve to fourteen inches apart in the rows.

it can be afforded, and in beet culture it will be found economical and convenient.

As soon as the seed are sown the ground should be rolled. This hastens germination. The best roller is a cast-iron one, in joints or sections. The roller should follow the lines made by the seed-sower as exactly as is possible.

The beet generally makes its appearance in about ten days after the seed is sown; but the time varies in accordance with the nature and condition of the soil and of the season. If the plant does not "show" in the usual time, seed must be examined in several different parts of the field, and if found generally to be alive, more time must be allowed for its germination. But if it be found that there is here and there a strong plant, while the rest come up irregularly, and examination of the seed in the vacant places proves them to be rotten, then it is to be considered,

1. Are there plants enough to give a fair crop?

2. Is the field in condition to allow seed to be sown in the vacant places?

3. Is there yet time to re-sow the whole piece?

WEEDING.

As soon as the plants are up, if weeds begin to appear, no time should be lost in setting the cultivator in motion to destroy them, and to stir the ground between the rows.

In Europe machines particularly designed for this purpose, as well as for other of the various requirements of this special culture, are in partial use. They will soon be brought here, and probably be improved upon by our skilful mechanics. But there are already cultivators in the west, that, with trifling modifications, would perform the required work admirably.

If the weeds should show thickly before the beet is up, and the lines made by the seed-sower are plainly visible, the cultivator may commence at once, for it is absolutely necessary, if good returns of beets, and also subsequent crops, are desired, that the fields should be kept entirely free from weeds.

In many parts of Europe the farmer not only runs his cultivator ("rasette a cheval") between the rows, but also across them, leaving his plants at the corners of squares eighteen inches apart each way, thus doing almost all his work with a horse cultivator. This implement sometimes operates on one, but oftener on three lines at once, and is drawn by a small horse, which is led by a boy.

The cultivator for one line does better work, but at a higher cost, than the three-line machine. There are two-horse cultivators in use, but it is difficult to employ a span of horses without injuring the crop. Many of these machines have a device attached that raises the leaves from the ground, and prevents their being injured. Others, also, have an attachment that "earths-up" the beet. The cost of these machines varies from five to thirty dollars. The one-horse machine, managed by a boy, will cultivate from three to four acres a day.

The use of the horse-cultivator *across* the lines is not recommended, as it leaves the plants too far apart in the lines. In some cases the hand hoe ("rasette a main") is used for both operations, and oftener still for cultivating across the lines. The "rasette a main" is mounted on low wheels, and is a species of thrust hoe and cultivator combined. The cultivator should not be run very deep upon its first passage, for fear of disturbing or covering the young plants.

In case the field is not cultivated across the lines either by the horse or hand rasette, it is necessary, as soon as cultivation between the lines has taken place, to thin out the beets, leaving single plants standing, from twelve to fourteen inches apart in the rows.* This can be done best when the ground is moist.

The ground should afterwards be loosened about the plant with a sharp, short-handled hoe, four or five inches in width, leaving the earth light and easily accessible to the fertilizing influences of the atmosphere.

In case there are vacant spaces in the lines, enough plants should be left in adjoining rows to furnish the means of filling the spaces by transplanting as soon as the beets are sufficiently large, which will generally be at the time of the second weeding.

Vacant spaces in the lines should be filled by transplanting. This can be done best when the beets are about one half or three fourths of an inch in diameter. A moist day should be selected, and the plants taken up with a spade, or, better, with a transplanting trowel, from those lines where thinning is required, and

* In thinning, particularly in dry weather, take a flat wooden knife with which to separate the plants and hold down the earth, while the beet to be removed is pulled up. If the earth is too dry to remove the plant easily, use a steel "dibble," with which the beet can be destroyed.

carried in a wooden tray to the spot where they are needed.

Pains should be taken to injure the beet as little as possible, and in replanting them to have the root kept straight, otherwise the matured plant will produce forked and misshapen roots.

Holes about five inches deep should be made at proper intervals for the reception of the plant, with a plug of hard wood, eight inches in length and an inch in diameter at one end, tapering gradually to the diameter of a quarter of an inch, when the end should be rounded off.

One careful workman should take up the beets and carry them to another, who will set them out. The latter workman takes a plant by the leaves with his left hand, and makes a perpendicular hole with the plug held in his right hand; he then withdraws the plug from the ground, and carefully inserts the plant in the hole, taking pains to keep the root perfectly straight. He holds it by the left hand, keeping the crown of the plant on a level with the surface of the ground; he then plunges the plug perpendicularly two or three times into the ground within an inch of the root, and crowds the earth against the root with the plug. He then places a little earth about the plant, and with both hands presses and settles the soil about the root. The earth is then dressed with the fingers about the plant, taking pains to leave the crown just even with the surface.

The long leaves are then pinched off, and the operations are completed.

The first workman should have a tray in which to

carry the plants he takes up. The second should also have one for the convenience of transporting the plants along the line. These trays should be two feet long and one foot wide, with rims three inches high on each side and on *one* end, with holes in the middle of the side rims to admit the fingers. One end is left open, so that the first workman can, with little injury to the plants, *slide* them from his tray into that of the second.

Transplanting may also be done when the plants are much larger, in which case it may be necessary to make the hole with the spade. In that case one workman thrusts his spade perpendicularly into the ground to a depth sufficient for the length of the root, and by a motion of his spade pries the ground to one side; another workman then inserts the root, holding it in its proper position; the first workman then withdraws his spade, and presses the earth against the plant with his foot. It is far better, however, to transplant when the beets are small.

There is also an instrument for transplanting, called a "deplantoir," in use in France, that moves the plant without retarding its growth in the least. It does the work perfectly in every respect, except that it does not do it expeditiously.

If the "spacing" of the plants is done by the passage of the cultivator across the lines, then the workmen must with their hands, or with the short hoe, loosen the earth about each plant, leaving but a single one standing at the corner of each square.

All weeds should be pulled up and left upon the surface between the rows, but not in piles, for they would obstruct the subsequent passage of the cultivator; whereas if they are spread, the cultivator will pass over them, and leave the intervals between the rows perfectly clean and smooth.

As soon after the first weeding as the ground becomes "baked" or "crusted," or as soon as the weeds again make their appearance, a second and deeper cultivation, and also a thorough weeding, should take place. The ground should at all times be kept pulverized, loose, open, and always entirely free from weeds. For this reason, as well as that the extirpated weeds may die more speedily, it is highly important that the weeding and cultivation should be done not only when the weather is hot, for then weeds are more easily killed, but also when the ground is dry, for it is at that time less likely to form a crust.

The only operations in beet culture suitable for wet weather are thinning and transplanting. Some hand labor is necessary, and frequent hoeings that break up the incrusted soil are of great benefit. Care should be taken to keep the hoes sharp, in order the more easily to cut off the weeds. There is a proverb in Germany that "the hoe is the gold of the beet."

The number of times that the beet should be weeded and cultivated is determined by the condition of the surface soil, and the existence of weeds. The weeds *must* be kept down. and the soil *must* be kept loose. Three weedings often suffice; if no more are required, so much the better. If six are needed, *they must be given.* The value of the crop demands this, and it *must* be done, and *well* done. The better it is done the first time, the less there will be necessary to do afterwards.

As soon as the plants take such full possession of the soil, that hoeing or cultivating cannot be done without damage to the leaves, then those operations must cease, for it is of great importance to preserve the foliage from injury. This will generally be the case early in July. The luxuriant growth of the plant then stifles the weeds, and, shading the ground, prevents its incrustation. The only care required after that time until harvest is to pull up such weeds as may have accidentally escaped the watchful eye of the farmer, and to cut off the flower-stalks of the few beets that give indications of producing seed.*

Both these operations must be strictly attended to, for the weed not only withdraws nourishment from the beet, but if permitted to mature, scatters seed that increase the farmer's subsequent labor; while the root of the beet that is permitted to "go to seed" contains not a particle of sugar.

It is a common but not universal practice in Europe to "hill," or to "earth up," the beet, and the method finds many advocates. The operation is performed principally with a species of small double mould-board plough, and is finished with the hoe. It is generally done between the second and third weedings. The practice seems to be a reasonable one, as it tends not only to make the soil light, and thus promotes the growth of the beet, but also causes its development beneath the soil, thus lessening the amount to be cut off of the neck at the time of harvest. Beets that have

* If the beet shows a tendency to go to seed while it is yet small, it should be pulled up; but if it is large, the flower-stalk should be cut off.

a tendency to grow out of the ground are improved in quality by earthing them up in July.

Harvesting the Beet.

The maturity of the beet is marked by unmistakable signs. The leaves of the plants, instead of looking green, thrifty, and vigorous, begin to assume a yellowish tinge, to wither and drop off. This period varies with the climate, the season, and also with the method of cultivating and of manuring. These indications are signals of the coming harvest, and the field must thenceforward be narrowly watched, calculation being made as to how much time will probably elapse before frost sets in; also as to the force attainable for harvesting the crop, and also as to the probability of rain. It is important that beets should be harvested before heavy frost, although they will, before being dug, bear a temperature of 22° to 24° without injury. Beets that are frozen should be left for eight or ten days before being dug, in which case they often recover from the effect of the frost; if they could be left longer, it would be still better. After being dug, the beet will bear a temperature of 28° without detriment. Heavy rains, after the foliage has withered and fallen, stimulate the production of new leaves at the expense of the sugar in the beet. This should be counteracted by harvesting the crop as speedily as possible; but the longer the beet stays in the ground without the risk of freezing or producing new leaves, the better for the manufacturer, and of course for the farmer, for their interests are identical.

Beets are generally ripe in France the last of September or first of October. In Illinois, by reason of the heat of our summers being more intense, they would ripen early in September. This is a great advantage to the manufacturer, as he can begin to work nearly a month sooner than is done in Europe, and operate upon ripe beets; while in Europe, the manufacturer, if he has a large crop to consume, begins in September, but has unripe beets, that do not contain their full proportion of sugar.

If a portion of the beets are to be taken at once to the factory, and the rest kept in pits for future working, then those that are ripest should be selected for the pits, and of the remainder the ripest should be first dug for immediate use at the factory. If the beets are all to be put into pits, then the least ripe, and also those grown on the richest ground, should be kept separate, and delivered first to the mill when they are required. The reasons for these rules are, that ripe beets keep better than unripe ones, and that beets grown on rich ground are more watery, and consequently do not keep so well as those grown on poorer soil.

Beets may be dug with a spade, fork, or common plough. They are generally taken from the ground in Europe with what is called an "arracheur," which is a sort of plough with a share shaped like a cone, the section of which is an oval somewhat flattened on the lower side, about three feet in length, seven or eight inches in diameter, and tapering to a blunted point. It is drawn by two horses, and will dig from one and a half to one and three fourths acres of beets per day in excellent condition. The operations of the spade or fork are too tedious and costly to be employed in this country, and the common plough injures a great many of the beets, thereby promoting their decay in the pits.

In harvesting the beet, it is advisable, chiefly for two reasons, to select dry weather and a dry state of the soil. If the weather immediately preceding harvest is very wet, not only is the beet rendered more watery, and the percentage of sugar contained in it less, — which of course is a disadvantage for the manufacturer, — but the beets will not keep so well in the pits. They are also more susceptible to the action of frost; for the richer the beet is in sugar, the better it will keep, and the less likely is it to freeze.

If the ground is wet, the earth also adheres more closely to the roots, and they are neither so easily dug nor so easily cleansed of the adhering soil. When the ground is wet and the extracted roots are very dirty, they must be gently knocked together to free them from the superabundant soil, but not with such force as to bruise them. Roots keep better when some soil adheres to them; but too much induces vegetation in the pits, which destroys the sugar.

When the roots are thrown out by the "arracheur," women and children place the beets from two rows side by side upon the ground, all lying in the same direction, with their leaves on one side and their roots on the other. This is for the convenience of the workman who cuts off the leaves. If the beets are properly placed, his labors are lessened, and he is not obliged to touch the beets with his hands. It takes but little extra labor, and that of women and children, to place them properly and to "double up the rows," — that is, to place in one line the beets from two rows, — while it not only hastens the labors of the cutter, but also facilitates the subsequent operation of throwing into pits or putting into wagons. Pains should also be taken to have two "doubled rows" come together, in order to allow the passage between the rows of extracted beets of a wagon, into which they can be loaded from both sides. This can be done in the following manner: rows one and two should, when dug, be laid on the ground occupied by row one; rows three and four on row occupied by four; rows five and six on row five; rows seven and eight on row eight; and so on.

After the beets are placed in lines, the leaves are cut off. For this operation several different methods are employed. In some instances the work is done by women and children, who use either a large knife with a curved point, like a pruning-knife, or a straight knife, with a blade about a foot long and an inch and a half wide. In other cases it is done by a man either with a spade or with an instrument shaped like a sod-cutter, with a handle about four feet long. This latter instrument is the best. Whichever is used, it must be kept sharp, not only to render the work easier, but also to prevent bruising the beet, which hastens its decomposition.

If the beet is of the right kind, and has been properly cultivated, so that the root has not pushed above the surface, it will only be necessary to cut off the foliage, just shaving the crown of the plant, so that the leaves fall separated; but if the root, for any reason, has grown much above the surface, then a portion of the green neck, which has been exposed to the air, should be sliced off with the leaves. Mutilation of the beet must be avoided, for every wound not only hastens decay in the pits, but even a slight exposure to the air induces fermentation at the wounded part, which somewhat lessens the production of crystallized sugar; therefore it is advisable to cultivate the plant so that it will not be necessary to cut off any part of the neck. After the leaves are cut off, the beets may be either put at once into the pits or silos, transported to the factory, or thrown into small piles. If the latter course is adopted, the piles should not be made more than two and a half feet high, and should be covered at once with the leaves as a safeguard against frost, and to exclude them from the unfavorable influence of light and air, which causes them to wither and become flaccid, and tends to promote decay.

The treatment after harvest is of the greatest importance: upon it depends the ultimate value of the crop, which may otherwise prove a total loss.

Preservation of Beets.

The methods of preserving beets are various. In some parts of Europe they are kept upon the surface of the ground, and in immense solid piles, covering acres of land to a uniform depth of about six feet. I have even seen them between nine and ten feet deep. In other cases they are placed on the ground in piles ten or twelve feet wide at the base, five feet high, and of any desired length, with the sides of the pile gradually converging as they approach the requisite height.

and with the top rounded so as to shed water. In some cases these piles are ventilated, as will be described hereafter, and in other cases they are solid. In the opinion of many, piles should not contain more than five tons, and should be thoroughly ventilated; on the other hand, I have seen more than 10,000 tons in a pile without any ventilation, and the beets came out in perfect condition. Some people preserve them in silos or pits of various sizes, ventilated, in some instances, and in others filled solid; in some sections the piles are conical.

In France a patent has been taken for the preservation of beets by the mechanical introduction of a current of cool air through ventilators that traverse the piles. Preservation in cellars is not practicable on a large scale, neither do the beets keep so well as those in pits or piles.

The best method of preserving the beet is to keep it continually frozen; for freezing not only does not injure its saccharine properties, but it facilitates the extraction of sugar, probably because frost ruptures the sap-vessels more completely than it is possible to do mechanically. The trouble of frost in Europe is, that a frozen beet, when it thaws, quickly becomes rotten, and it is impossible, in their climate, to *keep* them frozen; consequently frozen beets require to be worked at once, or decomposition takes place. In my judgment, beets may be frozen in Illinois in November, and by protecting them with straw from the rays of the sun, may be kept frozen until March.

As it is impossible for the manfacturer upon a large scale to take the whole crop at once, the usual method in Europe is to contract with each farmer for the delivery of his beets throughout the season. A portion is required each day, and is drawn to the mill for immediate consumption, if the beets are near to the manufactory. There are also provisions made for the storage of a large amount in the yards of the factory, and piles are also made on the road-side of adjacent fields. These supplies are drawn at the time of harvest, and are kept as a reserve for bad weather, or when, from any cause, the daily supply from the farmers is not sufficient. But if the factory is far from the fields where the beets are raised, the better course is to store the roots on the field, and deliver them as required; for the beets are injured by long transportation, and do not keep well. When the manufacturer has received all that he can take care of, the farmer preserves the remainder for delivery throughout the fall and winter.

In whatever way they are stored for preservation, it will be necessary to place all the outside beets in a perfectly symmetrical wall, gradually inclining towards the centre of the pile. For this purpose the beets are placed one by one, with their crowns out and the roots in. The rest may be thrown promiscuously into the interior of the pile. The sooner the beet is put into pits or piles after being dug, the better. In preserving beets, they must be kept from excessive moisture, prevented from heating, maintained at an even temperature, and be easily accessible in wet and freezing weather. In selecting places for their preservation, dry land that affords natural drainage should be chosen, and in close proximity to a road or highway, in order the better to keep them from excessive moisture, to permit their easy and frequent examination, and their more expeditious and economical transportation, without trampling upon and injuring the ploughed fields. If the piles are in the middle of the fields, and the ground is wet, more time and greater power will be required to draw out the beets than if they are at the road-side. I shall describe the method of preserving in "silos," generally employed in Europe, remarking again, that the size of these silos varies in accordance with the different ideas of cultivators.

Preservation in Silos or Pits.

A pit is dug in dry soil, from twenty to twenty-four inches deep, ten to twelve feet wide, and of any convenient length; the bottom rises a little at the centre. If the pit is perfectly dry, it will not be necessary to put anything on the bottom; but if it inclines to moisture, then it would be advisable to give it a coating of dry sand, and to make it sufficiently wide to have a ditch one foot wide around the pile of beets. This ditch should be five or six inches deeper than the bottom of the pile, and so arranged as to afford drainage for any water that might otherwise remain in the pits.

The roots are then put promiscouusly into the centre of the pit, and a symmetrical wall of beets, laid with the crowns out, at one end and at both sides. This wall must incline regularly towards the centre, at the rate of about one foot in three, care being taken to have the sides of the pile perfectly straight and even. When the pile has been carried up to the requisite height, or seven to eight feet from the bottom, and the beets on the upper surface smoothly and regularly laid, a portion of the earth that was taken from the pit must be thrown against the pile, and a wall of earth be built around the beets two and one half to three feet thick at the base, and gradually diminishing in thickness as the summit is attained. The thickness of this wall depends upon the climate and the soil: if the latter is very light and sandy, a greater thickness will be required than if it were heavier and of greater consistence. The top of the pile, for a width of three or four feet, is not covered with earth until the weather becomes cooler. This open space, however, is protected with six or eight inches of straw, which is kept in its place by boughs or sticks. It is better not to put the whole of the earth about the beets at once, but to cover them with only half the quantity at first, increasing the thickness of the covering as the season progresses. As the period for strong frost approaches, the straw covering on the top should be replaced by earth, the outside of the pile beaten smooth with a spade, and put in condition to remain through the winter. A transverse section of the finished pile resembles a haycock in form.

The end of the pile from which the beets are first to be taken, should be coated with three or four feet of straw, firmly secured with boards, so that access to the beets may be easily obtained when the ground is frozen hard.

In putting the beets into silos or pits, great pains must be taken to have all the beets in the pile of one condition; that is, the beets that will keep best should be put in one pile; those which will not keep so well should be put in another; those which are at all injured or bruised, if they cannot be used at once at the mill, should be put into a third; and so on, taking care to remember the character of the contents of each pile, so that those least likely to keep may be first delivered. This has already been referred to on page 107, but is of so great importance that I again allude to it.

Preservation in Piles.

Beets may be preserved in piles upon the surface of the ground in precisely the same manner as has been described above for their preservation in pits, with this exception, that the depth of beets should not exceed five or six feet. In both cases a shallow ditch should surround the pile and silo or pit for the purpose of drainage. In some cases a layer of beets a foot thick is covered with two inches of earth, and then another layer of beets, and so on until the pile is completed: this is a good but expensive process. In all cases the piles should be repeatedly examined, and all cracks and chinks in the covering of earth repaired at once.

Ventilation of Beets.

Some persons consider it of the first importance to ventilate beets, both those in piles and in pits; but I have seen such vast quantities kept in fine condition until the 15th of February, stored without ventilation, in the comparatively warm climate of France, that I doubt its necessity. When ventilation is practised, it is sometimes effected by placing in the centre of the pile, at distances of twelve to fifteen feet, chimneys two or three inches square, made of rough boards. These chimneys extend from the bottom to the top of the pile. In some cases a bundle of twigs five or six inches in diameter, and in others a pole wrapped loosely around with straw, takes the place of the chimney. Sometimes these chimneys rest upon the top of triangular frames or ventilators. These are made of a piece of board, perhaps a foot in width, and another narrow strip, say of scantling. Laths or short narrow strips of wood are then nailed upon the board and scantling, in such manner as to form a triangular frame, like the roof of a house, the board serving as the floor, the scantling as the ridge-pole, and the laths as the rafters.

These frames are placed end to end upon the ground, running longitudinally in the centre of the proposed pile, which is then placed about them in the same manner as described for the ordinary piles; the chimney is placed in the centre, and is connected with the ventilators, as has been described. Every twenty or thirty feet a frame also runs *across* the pile. The mouths of these ventilators come to the outside of the completed pile, and are stuffed and completely protected with straw, which can easily be removed, and by which the supply of air can be regulated. It is best, if possible, to have the piles and silos run north and south, having the end to be first opened facing the south. By this arrangement it is easier to protect the pile effectually, with earth and straw, from the influence of cold north winds; while the end which is to be opened, being on the south, is warmer and better protected.

Method of Preservation in Saxony.

A method of preserving beets prevails in Saxony that seems an admirable one, and well adapted to existing conditions in Illinois, where straw is superabundant and comparatively without value.

To facilitate the explanation I present the following diagram: —

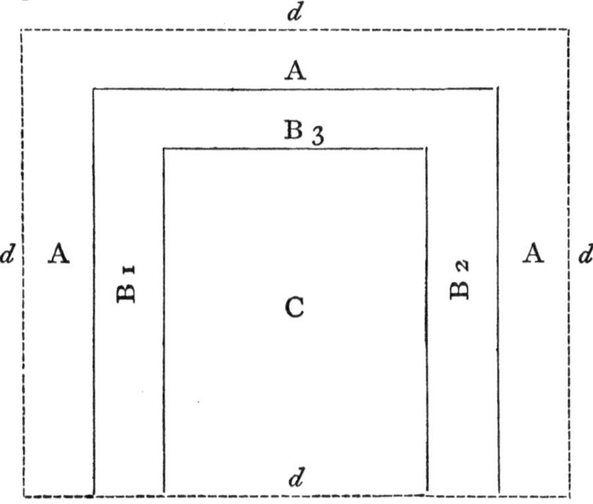

B 1, B 2, and B 3 are trenches, six feet broad and two feet deep, to be used as silos or pits, made chiefly with the plough on three sides of a parallelogram. The trench B 3 is fifty-two feet long, and the space C between the trenches has a breadth of forty feet and a length as great as may be needed. Storage is commenced by building piles in the silo B 3, in the manner described on page 113, beginning on the end of the parallelogram, and at the same time in the trenches B 1 and B 2. Line A is the outer edge of the pit. The depth of the beets in the piles should not exceed six feet, two of which being beneath the surface, gives a height of four feet above the ground. As the work progresses, the outside should be covered with earth and the inside with straw.

When the end is finished, and the two sides have been extended to a length of twelve or fifteen feet, the straw in the interior is removed, and other beets thrown promiscuously into space C, against the wall of beets in the trenches. The beets are eventually piled up to the level of the top of the wall; but in the early harvest, before the weather gets cold, it would be better to pile them only two feet deep, and put in the rest later in the season. The beets on the top, when the pile is finished, require to be carefully placed with their crowns on the outside and their roots extending into the pile. The pits are covered, as soon as they are finished, with straw, with which the inner part of the walls are also kept constantly protected.

Beets should not be left uncovered any longer than is absolutely necessary, from the time they are dug until they are consumed in the factory. As the weather grows cooler, the straw should be removed from the top of the pile, and a layer six inches thick of earth, or of short stable manure, spread, thoroughly smoothed, and rendered as compact as possible, upon the top of the pile. This layer may be succeeded, still later in the season, by a second or third layer, as circumstances require. When sufficient thickness has been obtained, the whole may be covered with straw. Separate compartments may be made with walls of earth in the space C, to separate different qualities of beets. The ends of the side walls in the silos must be covered with earth, like the rest of the outside; and when the whole crop has been harvested, the last of the beets must be employed to build a wall in the usual manner across space C. It will be necessary, in preserving beets in this way, to have a greater amount of earth for covering than the pits furnish; and in ploughing to procure it, furrows should not be run nearer than line *d*, say within three feet of the pile, lest the walls of beets should be disturbed. The advantages of this method are, that it allows the farmer to store large quantities safely on spots the most conveniently located. It also facilitates the daily opening of the pile, when the beets are to be carried to the factory; for the entrance is small compared with the size of the pile, and can be easily protected by straw, which it requires but little time to remove. It also saves, to a considerable extent, the comparatively tedious and costly process of building the walls which are required, when the smaller and consequently more numerous piles are constructed.

Method of Preserving Roots in Massachusetts.

I annex the instructions given for the preservation of root crops in Flint's "Agriculture of Massachusetts."

"Dig a pit six feet wide, ten or fifteen feet long, and eighteen inches deep. Pile the roots as steep and high as the base will carry and keep them. Cover the heap with a layer of straw six inches thick, and follow with a covering of earth six inches deep, patting it down, so that the rains shall not furrow it. Set one or more tile ventilators loosely filled with straw. In covering the heap, throw up the earth so as to leave a ditch around it about two feet from the base line, being sure to so construct it as to drain the water away. Cover the heap with an additional six inches of earth as late as the season will allow. Heaps of roots, however stored, must be properly ventilated. Vegetable matter is invariably decomposed by heat; hence the frequent loss invariably resulting from a want of care in storing them. Let them be kept at as low a temperature as possible above freezing point."

SEED.

The saving of seed is a matter of the greatest consequence, in connection with the production of the sugar beet. In the infancy of beet-sugar manufacture the ordinary forage beets, such as the red mangel-wurzel, that contains only five to six per cent. of sugar, and often less, were generally employed; * but varieties far richer were gradually introduced, and by judicious selections and crosses of different varieties, the character of the plant has been improved, and its saccharine properties largely increased. Experiments have

* This accounts, in some measure, for the low percentage of yield, and also for the high cost of sugar in former days; for the expenses were greater to work the poor beets, and less sugar was obtained, than is now done. In 1840 it required eighteen tons of beets to make a ton of sugar in the Zollverein. It now requires less than twelve tons.

conclusively shown, that seeds from beets rich in sugar, produce richer beets than are raised from the seed of those poorer in saccharine matter.

Beets containing sixteen per cent. of sugar are not rare, and in one instance twenty-one per cent. was found in a variety produced by M. Vilmorin, near Paris. There is no reason to believe that the saccharine quality of the beet has yet attained its complete development.

The German method of selecting the white Silesian beets to bear seed the succeeding year is as follows: —

They are chosen, not from the piles after they are gathered, but while they are still standing in the field rows. Medium-sized beets, grown in moderately rich soil, are preferred to those grown in land very highly manured. Plants should be selected whose roots, growing entirely beneath the surface, are shaped like a pear, and not like a turnip; whose crown is single, and presents no cavity; the longitudinal indentations on whose main root incline to a spiral rather than a straight direction; whose foliage is not too luxuriant, but, standing close together, grows low to the ground in form like a large plate; and the color of whose leaves is not tinged, spotted, nor fringed with red, but of a clear, bright green.

If varieties other than the white Silesian are used, then the properties to be sought for in the plant for future seed-bearing, should be those which most nearly approach perfection in the given variety.

The richness of a beet, either in saccharine, saline, or alkaline constituents, is determined by its specific gravity. But as the saline and alkaline properties are, to a certain extent, developed under different circumstances from those which produce the highest saccharine qualities, and as it is well known under what circumstances the one or the other properties are most fully developed, it is necessary to choose good seed-beets from those growing under conditions most favorable for the production of sugar, and from these to select those having the greatest specific gravity.

A soil rather sandy, and not too highly manured with stable manure, although it does not yield such heavy crops as one more fertile, nevertheless produces a beet that not only ripens earlier, but is also richer in sugar, comparatively free from saline and alkaline elements, and well suited for seed.

The salts in stable manure are readily absorbed by the beet; consequently the best course to take, in order to secure good seed-beets, is to sow the seed the preceding year on a part of the field that has not been manured for two or three years, and is best adapted by nature to the purpose. Bone-dust, however, may be used with advantage in the drill as a fertilizer. From the time the plant first makes its appearance, the cultivation should be most thorough. When the beets are ripening, select those having the qualities described, and mark them to be dug when fully ripe. When this period arrives, the roots are very carefully taken up, the extreme end of the tap-root removed, the leaves cut off with a sharp knife to within an inch of the crown, instead of close, as in the case of those to be used in the factory. A trench in a dry, well-drained soil, and in a sheltered spot, is then dug two feet deep, five or six feet broad, and of the requisite length. The beets are carefully laid side by side in the trench, layer upon layer. Between each layer just sufficient soil is spread to separate the rows of beets. As soon as the trench is full, the beets are laid so that the sides of the pile converge rapidly, assuming the form of an angular roof, the top of which is three feet above the surface of the ground. Earth is then put on to a thickness of eight inches, which is to be increased as the weather becomes colder.

As early in the following spring as the soil is in suitable condition, a piece of land, sheltered from the wind, and that was deeply ploughed in fall, is selected for the plantation of the seed-beets. Land should be selected for this purpose which is as distant as possible from other beets, in order to prevent the plants, when in flower, from being "crossed" by other varieties.

Deep furrows are made three feet apart, and holes are dug two feet apart in the furrows, of ample size for the reception of the root; the earth in each hole is made mellow with the spade, and two handfuls of bone-dust are incorporated with the soil. The beets are placed perpendicularly in the holes, without being bent, and the earth gradually put in and pressed about them with the hand. The crowns must be kept just below the surface. After the roots have been set out, and the earth thoroughly pressed against them with the foot, the ground must be dressed with the hoe, and one inch of earth, with a handful of bone-dust, placed on the crown of each plant, to protect it from frosts.

As soon as the beets are up, thorough cultivation and weeding must be persistently followed. The beets should be earthed up with the double mould-board plough and the hoe, the poorest and weakest flower-stalks removed, and as the seed begins to form, the tips of the stalks should be pinched off. Harvesting is done before the extremities of the seed-bearing branches turn brown. The stalks are cut off near to the ground, bound into small sheaves, containing eight or ten stalks, and kept until they are dry in a sunny and airy place. As soon as the stalks are well dried, the seed is thrashed out, dry, hot weather being more favorable for the operation. It is then winnowed and spread two or three inches deep on a dry and sunny spot, and occasionally stirred. When perfectly dry, it is put into sacks, not over a foot wide, and two and a half feet long, with labels attached, to describe the kind of the seed and the date of its production. The sacks are then suspended by cords in a dry, airy loft, in such manner that they do not touch each other, and are inaccessible to rats, which are very fond of the seed. Seed thus saved retains its germinating power for several years. In fact, seed only a year old should not be sown, as it produces beets more liable to "go to seed" than those obtained from old seed.

In some parts of Germany, after the seed-beets are taken from the pits in which they have been kept through the winter, they are subjected to a test, by which those only having the greatest specific gravity are retained for planting. The beets are all thrown into water, and the earth carefully washed from them; those which float are rejected, and the rest are reserved for the next test. Four or five vessels containing brine of different degrees of strength are then prepared, and the beets are placed one after another in the weakest brine. Those are rejected which float, and the remainder are subjected to the test in the next strongest brine, and so on, until those only are planted which sink in the brine that is strongest.

M. Vilmorin, the great seedsman of France, selects his seed-beets by making an accurate philosophical test of the density of the juice of each beet. For this purpose, with a sharp punch like an apple-corer, he cuts a piece out of the middle of the beet, punching it out with a wooden plug fitted to the aperture. This piece of beet he rasps, presses, and then filters its juice through a linen cloth into a "prover," in which, with the densimeter and aerometer, he ascertains its exact density. He retains only, beets of a certain standard of density. The holes in these are filled with sand, and they are planted in the usual manner.

A custom, borrowed from the Chinese, prevails in some parts of France, of making, before planting, three or four shallow, longitudinal cuts on the side of the seed-beets (beginning an inch or two below the crown), which open during vegetation. The theory is, that roots are thrown out from these cuts, and the beet is thereby enabled to draw sustenance from a more extended area, throwing up a stouter flower-stalk, less likely to be influenced by the wind, and producing better and more abundant seed.

Too much pains cannot be taken to plant the best seed, for beets vary so much in saccharine richness in districts where little attention is paid by farmers to selecting the best kind, that many manufacturers test the beets before purchase, and pay according to quality, not quantity, some beets being really worth twice as much as others.

Crops very rich in sugar are not so large as those of a poorer quality.

Where beets are sold by the ton, and not by degree of richness, those containing twelve to twelve and a half per cent. afford the density upon which the interests of the farmer and manufacturer can best be united.

In the infancy of the industry in this country, we shall be compelled to import seed. All varieties, and of the best qualities, not only of beets, but of all other plants, may be obtained, with certainty of being true to description, from Vilmorin, Andrieux & Co., of Paris.

Manures.

For the profitable production of beets in Europe, the liberal use of fertilizers is a necessity. The virgin soils of the west may not absolutely require it, in order to secure good crops, but there is no doubt that productiveness can be increased by the judicious use of manures; and it is quite certain that the time will soon come when it will be absolutely necessary. The best fertilizers to produce large crops of beet, are human ordure, and that of horses, cattle, and sheep. The urine, and all liquid manures, should be saved, because they are richer in fertilizing properties, and assimilate more readily with plants than the solid portions of dung. There is, however, this objection to the use of all those fertilizers which are rich in salts, that, when freshly applied to a crop of beets, they cause the latter to flourish vigorously and give large returns, but the presence of salts is prejudicial to the economical extraction of sugar, and the roots abound in saline elements that are absorbed from the manures. In Germany, where beets are taxed, and quality is of more importance than quantity, the beet is not sown on freshly manured land, but on soil that has not been enriched for one or two years. In France, on the contrary, where the sugar is taxed, and the object of the farmer is to get large crops, the beet is sown on soil highly manured the preceding fall. The consequences of these two systems are, that the crops in France, although considerably heavier than those of Germany, do not possess as rich saccharine properties. The German beet is more than one per cent. richer than the French, owing to the facts, that it is by nature richer, that it grows in a colder climate, and, following the law of latitudes, secretes more sugar, while, at the same time, its growth not being so much stimulated by manures, the same amount of sugar is diffused through a smaller space. The average production of sugar on an acre of land in the two countries is about the same.

The ordure of cattle produces cleaner, smoother, and handsomer roots, containing fewer salts, than that of men, horses, sheep, or swine. Indeed, there is quite a general prejudice against the use of sheep and hog manure on beets in Europe, whether well or ill founded I am not able to say. Many farmers fatten sheep on the pulp of beets, upon which they thrive admirably; but the method is not universally approved even by those who practise it. They say that sheep manure is bad for the beet, but the disadvantage is in a measure compensated for, when the pulp is fed to sheep, by the excellent quality of mutton produced.

Well-rotted, strawy manure is preferable to that without straw. If applied to the beet, without any preceding crop, it should be done in the fall, as directed on page 92; if employed in spring, it should be thoroughly "worked over," and made as fine as possible. Stable manures may be advantageously composted with muck, with wood or coal ashes, or with the young beets which have been thinned out, if they are not all required for stock. Muck may be composted with lime, or ashes, either of wood or coal. The refuse of the sugar manufactory furnishes great quantities of fertilizing materials, most of which are of the very highest value. The earthy refuse of the wash-house, where the beets are cleansed before rasping; the little roots and fibres; the decayed portions of such beets as it may be necessary to trim; the scum of defecation; the incrustations of the boilers, reservoirs, and cisterns; the worn-out sacs; the waste and exhausted bone-black; the ashes from the boilers; and the exhausted lime of defecation, are of great value. They are all sources of revenue to the European manufacturer, and I have even seen mill-owners, besieged by applicants for the privilege of buying the mud accumulated in their factory yards from soil that fell off the wheels of wagons used in transporting beets.

The scums and incrustations, the lime and the bone-black, should be mixed thoroughly together with an equal quantity of fine soil, and applied either broadcast in the spring before harrowing, or sprinkled about the plants at the time of cultivation. It is also an excellent compost for seed-beets, and can be put into the holes and incorporated with the soil at the time the beets are "set out."

Bone-dust and superphosphate of lime, particularly the former, are excellent fertilizers. Peruvian guano is a powerful stimulant, but its effects, when used alone, are not favorable; it is better when mixed with the two previously named manures in the proportions of one of guano to two, or even three, of bone-dust or superphosphate of lime. These manures should be thoroughly mixed, and kept from the air for one week before they are used.

Such of the beet leaves cut off in the fields as are not wanted for stock, when spread upon the ground and ploughed in while green, furnish an excellent manure, equal, if all are left, to six or eight loads of stable manure per acre. Linseed oil cake powdered, and sown broadcast before harrowing at the rate of half a ton or a ton to an acre, or sprinkled about the plants at the time of cultivation, is an excellent fertilizer. Bone-dust and wood-ashes, or bone-dust, ashes, and lime, in equal proportions, are excellent. Lime from gas-houses, thoroughly mixed with stable manure, makes an excellent compost for the beet. Chloride of sodium or common salt, which on some soils and for certain crops makes a good compost, is very unfavorable for the sugar beet, unless mixed with certain other materials in the form of an artificial fertilizer, which I shall hereafter describe, and which has produced great results.

An artificial manure, manufactured by Emil Güssefeld, of Hamburg, by treating the guano from Baker's Island with sulphuric acid, is in high repute in Germany. It is called Güssefeld's superphosphate of Baker's guano, and is thus composed: —

Phosphoric acid,	19.9
Magnesia and lime,	17.3
Sulphate of lime,	42.1
Water,	16.2
Organic substances,	2.9
Alkaline salts,	1.
Other matters,	.6
	100.

Three adjoining pieces of land, containing .63 of an acre each, were cultivated with beets in the following manner, and showed the given results. For the convenience of the reader the table is made on a basis of one entire acre: —

Number of the fields.	Amount of superphosphate, lbs.	Peru guano, lbs.	Quantity of beets, lbs.	Quantity of beets, tons.	Contents of sugar, per cent.
1	177		29.524	13¼	12.49
2	355	177	29.524	13¼	13.23
3	531	265	30.158	13½	13.62

All manures having a basis of potassa, or that contain soluble phosphates, are of the highest value in the culture of beets.

It is said that the use of sulphate of potash, as manure, increases their saccharine contents two or three per cent. Instances are recorded of beets so fertilized containing twenty per cent. of sugar.

Güssefeld also makes an artificial manure containing fourteen per cent. of potash and thirteen per cent. phosphoric acid. Experiments go to show that it increases the yield of beets over that obtained on unfertilized land from twenty-five to thirty per cent., and the percentage of sugar about one half per cent. An artificial manure, containing seventeen per cent. of soluble phosphates, made by Gils & Co., of Antwerp, costing $48 a ton of 2,240 pounds, was applied to land in Saxony at the rate of 325 pounds per acre, or at a cost for manure of about seven dollars.

The following were the results, as compared with a field precisely similar and well manured with stable manure: — Crop on an acre, with stable manure, 31,064 pounds, or 13.87 tons; with artificial guano, 48,741 pounds, or 21.76 tons. Difference in favor of guano, 17,677 pounds, or 7.89 tons.

In seventeen cases recorded in Saxony, fields manured with Peruvian guano, mixed with this fertilizer in the proportion of two of the former to three of the latter, produced, as compared with unmanured land of equal original condition, an increased crop of 3⅓ tons per acre.

It is used in Saxony at the rate of over 12,000 tons a year.

A Mr. Frank, of Stassfurt, in Prussia, near Magdeburg, has compounded an artificial manure from the refuse rock salt of the mines in his neighborhood.

This manure, costing at Stassfurt about forty dollars per ton, contains the following constituents in the proportions set forth: —

Sulphate of potash, . . .	18	to	20	per cent.
" " magnesia, . .	18	"	20	"
" " lime, ...	3	"	5	"
Chloride of sodium, . . .	40	"	42	"
" " magnesium, . .	2	"	3	"
Magnesia, ...	2	"	3	"
Water and sand, ...	17	"	7	"
	100		100	

This manure is spread upon the land at the rate of 175 to 350 pounds per acre, either in fall or spring, and ploughed in; or it may be mixed with guano in the proportions of two of the former to three of the latter; or it may be mixed and thoroughly incorporated with stable manure. Experiments were tried with it in Waldau, Prussia, in 1864, on a large scale, no less than 500 tons of the manure having been employed, at the rate of 180 to 350 lbs. to an acre. Fields containing from twenty-five to fifty acres were chosen for the trial: these were manured in the usual manner; were then divided into equal parts, and the Stassfurt manure added to one of the parts. The greatest pains were taken to give it a fair test, and the following results were obtained. The yield of beets slightly exceeded that on other fields. To give an idea of the astonishing excess of sugar contained in beets produced with the Stassfurt manure, the following table is submitted: —

Fields.		Stassfurt manure per acre, cwts.	Sugar in juice with Stassfurt manure, per cent.		Sugar in juice without Stassfurt manure, per cent.	
1	½ ½	none 1.	14.04	Average, 14.95	12.42	Average, 13.23.
2	½ ½	none 1.	16.20		12.83	
3	½ ½	none 1.	14.51		13.17	
4	½ ½	none 2½.	15.63		14.43	
5	½ ½	none 1.	14.38		13.30	

The following also shows the effect of the Stassfurt fertilizer in other parts of Prussia: —

Applied at the rate of 533 pounds to an acre, it increased the quantity of sugar in the juice of the beets over those raised without the manure from 12.82 to 14.42 per cent. In another instance the increase was from 13.6 to 14.8 per cent.

The best method of employing it seemed to be with bone-dust or phosphate of lime, in the proportions of one of the former to three of either of the latter. Guano and Stassfurt manure, with bone-dust or phosphate of lime, in the proportions of one each of the former to three of either of the latter, have been applied with excellent results.

Stable manure alone will not supply the materials taken from the soil by crops. For this purpose artificial fertilizers are required. Chemistry not only teaches us of what materials these fertilizers should be composed, but also provides them. According to Hall and Ogston, English chemists, the amount of solid material removed from a field with every ton of beets is as follows: —

	Roots. lbs.	Leaves. lbs.
Potash,	4.99	7.86
Soda,	3.02	2.52
Lime,	.41	3.31
Magnesia,	.43	3.27
Oxide of Iron,	.12	.52
Phosphoric acid,	.66	1.94
Sulphuric acid,	.65	2.20
Chloride of soda,	5.29	12.82
Silica,	.54	.76
	16.11	35.20

Calculating the average yield at twenty tons to an acre, and assuming that the leaves, as well as the roots, are removed, there would be taken from each acre $1026\frac{20}{100}$ pounds of solid material.

In order to maintain the fertility of the soil, it will be necessary to return this amount to the land. Stable manure will not provide all the requisite materials, and the deficiency must be supplied with properly composed artificial fertilizers.

Barral says to French agriculturists, "Buy artificial manures, but above all increase your stable manure."

Rotation of Crops.

The necessity of a rotation of crops is too well established to be discussed — the only question is, What is the best succession? I am aware that some farmers, particularly at the west, proceed upon the theory, that the fertility of their land is inexhaustible, cultivating the same crop year after year upon the same soil, and in too many instances without manure. I have seen fields upon which corn had been raised for twenty-two successive years without manure — a folly even greater than Crespel records, when he states that he cultivated sugar beets for fifteen successive years on the same land. This method of farming — if it can be called farming — is pernicious and suicidal, and should never be copied. In many parts of Europe the system of rotation is biennial, namely, wheat and beets; but it is never adopted by the best cultivators, and is rapidly falling into disfavor. In some countries beets are raised on the same land twice, and in others three times, in five years. The triennial system is the one generally in use; but among the very best cultivators, beet is raised on the same soil only once in four, or even five, years.

I shall give the crops often employed in Europe in the triennial and quadrennial systems of rotation: —

Triennial System.

1st year,	oats manured,	Or oats,
2d "	beets,	beets manured,
3d "	wheat.	wheat.

Quadrennial System.

1st year,	wheat,	Or clover,
2d "	beets manured,	rye or oats,
3d "	barley or oats,	beets manured,
4th "	clover.	wheat.

Where wheat is not much cultivated, rye may take

its place. Potatoes, if well manured, or barley, may take the place of oats.

The beet is excellent to precede all grain crops. It is a good successor of potatoes well manured, or of corn, and especially of rye or oats. It is a good successor of tobacco.

It is a bad successor of clover; and worse still of turnips, carrots, or forage beets.

The quadrennial system of rotation permits quite a range in the selection of crops, and change is beneficial to the soil, and consequently to the crops. It would be desirable so to arrange the fields, that clover should not be raised on the same soil oftener than once in eight years.

Beet Pulp.

After the juice is expressed from the rasped beet, the dry pulp remaining is an admirable food for cattle, sheep, swine, or fowls, of which vast numbers are fed in the sugar-producing districts of Europe. The average amount of pulp is twenty per cent. of the original weight of beets, and it is almost a universal custom for farmers to contract with manufacturers to receive back in pulp twenty per cent. of the weight of beets furnished. For this the farmer pays two to two and a half dollars per ton. If the manufacturer has any pulp remaining after his contracts with the farmers are filled, he sells it to others at two dollars and seventy-five cents to three dollars per ton.

Repeated experiments have proved that for feeding stock, three tons of pulp are fully equal in value to one ton of the best hay. Cattle are very fond of it, and by its use are fattened for the market in one hundred days.

The method of feeding stock upon it, employed at Masny, by the Messrs. Fievet, the model farmers of France, was the following: —

Each ox was allowed daily
80 pounds of pulp,
5 " " chopped straw,
5 " " oil-cake.

Each cow was allowed daily
70 pounds of pulp,
5 " " chopped straw,
5 " " oil-cake.

Each sheep was allowed daily
6 pounds of pulp,
½ " " chopped straw,
½ " " oil-cake,
1 " " chaff.

They fattened in this manner 800 head of cattle and 3000 sheep every year.

The Messrs. Fievet recommended the use of chopped cornstalks and a small quantity of Indian meal for the Western United States.

Chaptal says of the pulp, "This food is almost dry; it has not the disadvantages of grasses or roots, nor of dry forage. It does not ferment, and is not laxative, like the former, nor does it heat and produce constipation, like the latter. It contains almost all the nutritive principles of the beet."

In fact, water is the chief article taken from the

beet by rasping and pressing, and there still remains from four and a half to six and a half per cent. of sugar in the residuum, besides other nutritious matter.

Dombasle recommends it, especially for sheep, and also for milch cows, stating that the quantity, as well as the quality, of the milk, and the color of the butter, are much improved by its use.

M. Cail, the wealthy and enterprising owner of "La Briche," — a splendid farm in the department of Indre et Loire, — mixes his pulp with chopped straw, in the proportion of five sixths of the former to one sixth of the latter. To the oxen, for fattening, he gives 150 pounds of this mixture in the winter months; to milch cows, 110 pounds; and to working-cattle, from 100 to 150 pounds daily.

A liberal daily allowance for an ox is 75 pounds, for a cow 60 pounds, and for a sheep 6 pounds, with chopped straw, and a little oil-cake, or meal. Consequently, if a farmer raises 100 tons of beets, and takes back from the manufacturer 20 tons of pulp, he has the means of feeding, during the five months from the first of November to the first of April, 4 oxen, or 5 cows, or 50 sheep. The manufactory that consumes 24,000 tons of beets provides 4,800 tons of pulp, with which may be fed, for the five most costly months of the year, when there is no pasturage, 960 oxen, or 1,200 cows, or 12,000 sheep.

Preservation of the Pulp.

Beet pulp may be kept perfectly good for several years. I have seen at Masny cattle eat with avidity that which was two years old.

The method of preservation there adopted was to dig a ditch of any required length, eight or nine feet deep, and of the same width, in a soil so dry and hard that there was no danger of the sides crumbling.

The bottom of this ditch was a little lower on one side than on the other, to permit any water that might exude from the mass to settle in the lower part. The pulp was then packed and trodden solid into the ditch, raised one or two feet above the surface at the sides of the trench to allow for the settling of the mass, then built up into the form of a sharp roof, and the whole covered with one and one half to two feet of earth, beaten solid with the back of a spade.

Where the soil is not of a nature to allow the walls to stand safely, the pit is walled with bricks laid in cement.

Leaves.

The practice of plucking from the beets a portion of their leaves for feeding stock prevails in some districts, but it is entirely unadvisable. When it is done, the stripping begins in the month of August. Two or three leaves are taken from each plant, until a sufficient supply is obtained for the daily wants of the herd.

The reasons why the practice is a bad one are, that, the leaves having important functions to perform, the removal of the foliage impairs those functions. Nature also makes an effort to repair the loss, and new leaves form at the expense of sugar in the root. The period of maturity is also retarded; consequently the crop is less likely to keep well, beets perfectly ripened being more easily preserved than those which are less ripe.

The general custom is at the time of harvest to feed to the stock all the leaves they require, and to spread the remainder on the fields when they are cut off, and plough them in while yet fresh and green. In this way they serve an admirable purpose as manure.

But they are sometimes gathered and put in layers into trenches. Between each layer coarse salt is sprinkled; the pile is then covered with a layer of straw, and finally a thick coat of earth is added.

Leaves used as Fodder for Milch Cows.

The effect produced on milch cows by this food, and also the method adopted for preserving the leaves, are shown in the recorded experiments of Drs. Wels and Tod of Maiz-Blanco, in Moravia.

The experiment was made with six cows of the race of that country. Their conditions were as nearly similar in age, size, weight, yield of milk, and duration of milking, as could be desired for a fair test. For eight to twelve weeks they were fed daily with thirty-five pounds of beet pulp, five pounds of salted leaves, and six pounds of chopped barley straw. They gave regularly about the same quantity of milk in the aggregate and individually. After that time, commencing February 7, three of the cows, designated as A, received each forty pounds of pulp, six pounds barley straw and half an ounce of salt per day for four weeks. During the same time each of the three cows, designated as B, received daily twenty-six and two thirds pounds of pulp and six pounds barley straw; they also received daily, of salted leaves, in addition, for the first week, thirteen and one half pounds; for the second week, sixteen and one half pounds; for the third week, twenty pounds; and for the fourth week, twenty-six and one half pounds.

The result was, that the amount of milk given by the cows A fell gradually in the four weeks from an average of 25.78 pounds per day to twenty pounds per day, while the cows B increased their average daily production from $26\frac{9}{100}$ to $31\frac{40}{100}$ pounds.

During the next four weeks the cows A were fed in the same manner that the cows B had been, and the cows B were put upon the old diet of the cows A, with this exception, that they received, besides the daily allowance of forty pounds of pulp and six pounds of barley straw, an additional daily allowance of pulp to the extent of twenty-six and one half pounds the fifth week, twenty pounds the sixth week, sixteen and one half pounds the seventh, and thirteen and one half pounds the eighth week.

The result was, that the cows A, now fed on the leaves, gradually increased their average flow of milk, until, at the end of the second period of four weeks, the yield had risen from 20 to $29\frac{53}{100}$ pounds, or considerably more than that at the beginning of the experiment eight weeks before, while the daily yield of cows B, fed chiefly on pulp, fell from $31\frac{40}{100}$ to $21\frac{25}{100}$ pounds.

The following is a tabular statement of the result for the first four weeks, and of the yield of cows A for the last four weeks:—

Cows.	Quantity of pulp.	Quantity of straw.	Quantity of leaves.	Yield of milk, first four weeks.	Yield of milk, last four weeks.
	lbs.	lbs.	lbs.	lbs.	lbs.
A	3360	504	—	618.44	—
B	2240	504	1610	834.69	—
A	2240	504	1610	—	761.36

The daily yield of milk from cows A, when fed upon leaves, rose from 20 pounds to 23.75 in the fifth week, 26.87 in the sixth, 28.44 in the seventh, and 29.53 in the eighth.

The milk produced by cows,—

A in 1st and 2d week, on pulp, averaged 2.8% butter.
B " " on leaves, " 3.7 "
A in 3d and 4th week, on pulp, " 3. "
B " " on leaves, " 3.9 "
A in 5th and 6th week, " " 3.2 "
A 7th " 8th " " " 3.6 "

The experiment ended April 3d.

The conclusions are, that salted leaves can be preserved; that their use, in conjunction with pulp, increases the flow of milk and yield of butter; and that they are preferable to pulp alone.

The weight of leaves is about twenty per cent. of the weight of the roots; therefore a factory that consumes 24,000 tons of beets annually furnishes 4,800 tons of pulp, and the beets furnish 4,800 tons of leaves — an amount of fodder sufficient to feed nearly 2,000 oxen, or 2,500 cows, or 24,000 sheep for five months.

The following method is adopted for preserving leaves of the sugar beet: —

Ditches are dug and walled with brick, backed with clay, and laid in cement, so that the interior is nine feet long, seven feet wide, five feet deep, and perfectly impervious to water. A layer of leaves, three or four inches thick, is spread on the bottom; this is sprinkled with coarse salt; then a layer of chopped straw one inch thick; then another layer of leaves; and so on, until the reservoir is filled. These are all packed down as solid as possible, and are then covered with six or eight inches of long straw; the whole is protected with earth, or, better still, with boards, held in their places by stones. About 225 pounds of salt are required for each pit.

Advantages of Beet Culture to Farmers.

The introduction of beet sugar manufacture into the United States would be of great benefit to farmers. It would insure to them superior methods of agriculture, increased crops, more remunerative prices, home markets, and enhanced value of farms. It would create industry, and diversify labor, thereby increasing the general prosperity, intelligence, and happiness of the community. It would eventually reduce the prices of sugar, of bread, and of meat, and render the United States more independent of foreign countries.

APPENDIX.

Since the foregoing pages were written, considerable information has been acquired, which throws additional light upon the subject treated of in this volume, particularly in relation to the cost and quality of American beets.

As regards cost, the estimates of cultivators, based upon results on their own farms, vary from 75 cents to $3.75 per ton.*

P. T. Quinn, of Newark, New Jersey, manager of the farm of the late Professor Mapes, says, that after land has been "broken up" he can cultivate sugar beets at $16 per acre, in the best manner, "not letting a weed show itself," and obtain crops of from 25 to 30 tons per acre. The above cost covering every expense, including that of harvest.

J. C. Thompson, of Staten Island, says he has obtained 40 tons of sugar beets from an acre, and that he can certainly get 30 tons at a cost not exceeding $25, and by extra pains, could obtain 50 tons.

Emory Rider has raised 30 tons per acre, at Hackensack, New Jersey, and counts with certainty upon 20 tons, at a cost not exceeding $28, including "pitting" the beets.

Hon. Ezra Cornell, of Ithaca, New York, has raised 20 tons per acre, and is confident that all expenses cannot exceed $30, or at most $35 per acre. He believes that beets would be a profitable crop at $2.50 per ton.

Sugar beets have been cultivated on a large scale in Illinois, upon the farm of a wealthy land owner, at a cost

* As the ton is generally reckoned at 2000 pounds in the United States, the figures in this Appendix will be based on that weight, although in the preceding volume the ton was reckoned at 2240 pounds.

of $30 per acre, including breaking up the prairie. Crop on raw prairie ground 10 tons per acre; crop on improved land 15 tons per acre. He believes there is no difficulty, after two or three years of cultivation, in raising beets at $2 per ton.

The late William H. Belcher, of St. Louis, believed, as the result of very extended and particular inquiries and observation, that beet could be raised at $2 per ton.

Theodore Gennert, of Chatsworth, Illinois, who raised 400 acres of beets last year, stated that they cost less than $3 per ton, and believes that when the soil is fully subdued they can be raised a good deal cheaper.

Maurice A. Mot raised in 1862 ten acres of beets at Cherry Valley, near Newark, Ohio. The soil, with the exception of little more than an acre, was quite poor, and the crop very light on the poor land, but his beets cost him only $2.65 per ton. Several of the neighboring farmers offered to cultivate another crop of beets on the same ground the following year at $10 per acre.

Joseph Sullivan, of Columbus, Ohio, says, "I have no doubt that an average yield of 30 tons of beets per acre, upon good, suitable soil, moderately well cultivated, can be secured. Corn ground which produces sixty-five bushels per acre, may be easily made to produce 30 or 35 tons of beets."

The late John W. Massey, of Morris, Illinois, wrote in 1865, in relation to cost of cultivating beets, and their yield in his region, "that it would take a little more work per acre than corn, and probably less than potatoes. His experience of more than 20 years in Illinois led him to believe that the cost of cultivating an acre of beets would be about the same as for sorghum, say about $30, and that the crop was 15 to 20 tons per acre.

John W. Walsh, of Chicago, published in 1863 a pamphlet, "Observations on Beet Sugar and Sugar-beet Culture," in which he stated that 15 tons per acre was a fair yield, but 18 to 24 was not uncommon. That he had known of frequent instances of crops of 36, 39, and 42 tons, and even as high as 90 tons being produced on rich loams.

10

Mr. Sanford describes, in the *Genesee Farmer*, a crop grown by him of 63 tons per acre.

T. E. Payson, of Deer Island, Boston Harbor, raised 73 tons long red mangel-wurzel on an acre of land in 1866.

Dr. Lettson has raised in England 120 tons mangel-wurzel, tops and roots, on one acre. This is equivalent to about 96 tons of roots. Even this enormous yield has doubtless been surpassed in the same country by scientific culture, for the writer has heard descriptions of crops that he cannot now authenticate, of over 100 tons of roots per acre. The yield of sugar beets is usually about two thirds that of mangel-wurzel.

The average yield of sugar beets in France is over 20 tons per acre. It often rises to 50 tons, while instances of 60, 70, and even 90 tons are not uncommon.

In Germany the yield varies from 10 to 25 tons.

Figures, made by Mr. Walsh, to whom reference has just been made, indicate that in his judgment beets can be raised in the West for less than $2 per ton, and that at $3 per ton they would prove to be very profitable to the farmer. The average price paid by European manufacturers is less than $3 per ton.

The estimate of the Agricultural Department of the United States is, that they can be raised at a cost to sugar manufacturers of $2.60 per ton.

In the light of all this testimony, as well as that on pages 26 to 39 of this volume, together with the additional fact that beets were furnished by Western farmers in 1866, on contract, for $3.50 per ton, it is not deemed extravagant to assume that sugar manufacturers can be supplied with beets in the West at $4 per ton.

The next point to establish is the saccharine property of the beet of America as compared with that of Europe.

The average percentage of sugar in the French beet is 11½, and in the German beet 13. This is the result of many years of scientific culture, by which the original saccharine properties of the beet have been increased.

There have been hundreds of tests made in this country

within a short time, showing a range of from 8 to more than 17½ per cent. of sugar. These experiments were made for the most part upon poorly cultivated beets, grown for feeding to stock, and not for sugar making, consequently bulk, and not quality, was the desideratum. It is well known that a skillful use of manures will increase the saccharine properties of beets. None of the American beets tested were grown under the most favorable circumstances, yet the result of the tests is entirely satisfactory.

The following table, showing the results of a few tests, indicates the richness of American beets :

Kind of Beet.	Where raised.	Percentage of Sugar.
White Sugar, . . .	Hackensack, N. J., ...	$14\frac{4}{10}$.
" . . .	" "	$17\frac{8}{10}$.
" . . .	Roxbury, Mass., ...	$11\frac{2}{10}$.
" . . .	" "	$12\frac{6}{10}$.
" . . .	" "	$13\frac{1}{10}$.
" . . .	Dedham, "	$10\frac{2}{10}$.
" . . .	Shirley, "	$12\frac{6}{10}$.
" . . .	" "	$14\frac{8}{10}$.
" . . .	Deer Island, "	$10\frac{4}{10}$.
" . . .	Chatsworth, Illinois,	9 @ 14. average 12.
Red Top Mangel-Wurzel,	Dedham, Mass., ...	$9\frac{1}{10}$.
"	Deer Island, "	$10\frac{1}{10}$.
Green Top Mangel-Wurzel,	Morris, Illinois, ...	11.
"	"	$9\frac{4}{10}$.
Yellow Globe, . .	Deer Island, Mass., . . .	$8\frac{1}{10}$.

It will be seen by the above table, that the white sugar beet, exclusive of those raised at Chatsworth, averaged $12\frac{9}{10}$%. That those raised in Chatsworth averaged 12½%. The Mangel-Wurzel average $9\frac{9}{10}$%, and the Yellow Globe contained $8\frac{1}{10}$%.

These tests, made in 1866, as well as others made during the past six years, prove conclusively to the writer's mind that American beets are richer in sugar than the French. That they are as rich as those of Germany, and that by proper culture, their saccharine properties may be increased at least one per cent.

If, then, it has been proved that we can economically raise beets in the United States as rich or richer in sugar than those of Europe, it only remains, in order to insure independence of the rest of the world for our future supplies of sugar, to prove that sugar can be extracted from them at prices enabling the manufacturers to compete with foreign countries.

The Germania Sugar Company of Chatsworth, Illinois, to which reference is made on page 62, has solved the problem, and although the enterprise has not as yet proved an entire success, yet the causes of its partial and temporary failure are so clear, and so easy to avoid, that no person conversant with the facts can deny that the Company has proved that the manufacture of beet sugar is not only entirely practicable, but must inevitably be highly remunerative.

I annex a report, made to the Directors of the Illinois Central Railroad Company, by R. W. Bender, of New York, who visited Chatsworth in January, at the request of the Company, to inquire into the cause of the troubles which were encountered. Mr. Bender is a practical refiner of large experience, and his opinion on the subject of which he treats is conclusive with all who know him, for in his great desire not to mislead others, he is very conservative in the expression of his opinion, taking care to understate rather than to overstate his estimates of the practicability of manufacturing beet sugar in the United States.

To the Directors of the Illinois Central Railroad Company:

GENTLEMEN:

Having recently returned from a visit to the Beet Sugar Works at Chatsworth, Livingston Co., Illinois, I now report the result of my investigation of this subject.

The works are owned by the Germania Beet Sugar Company, located as above, and are under the management of the Messrs. Gennert, the original projectors of the enterprise.

They commenced operations for the season of 1866, by planting four hundred acres of land, mostly fresh prairie, from which they have raised a crop of more than four thousand tons of fine beets, at a cost, according to their estimate, of less than four dollars per ton.

The beets are of the White Silesian and Imperial varieties, and both have done well. At the time of harvest, Messrs. Gennert tested the roots from all parts of their farm, and found the juice to contain from nine to thirteen and a half per cent. of sugar, by the Soleil Polariscope. The average of all the tests showing twelve per cent.

This result is confirmed by the investigations made by direction of the Belcher Sugar Refining Company of St. Louis; the tests they made, showing an average of twelve per cent. of sugar in the juice, and in some cases as high as fourteen per cent.

At the time of my visit (Jan. 29th, 1867), I obtained what I considered fair samples of the beet roots, and found them to contain from nine to eleven per cent. of sugar (in the juice), with foreign substances amounting to about five per cent.; not a very undue proportion, considering the fact that the roots were principally grown in fresh prairie soil, and that the fall season was a wet one. My analysis of the juice fully confirmed the results obtained by the Messrs. Gennert and the Belcher Refining Company three months previously, at the time of harvest, and when the process of sugar manufacturing should have commenced.

The quality of beets, shown by the foregoing experiments, would yield 7⅛ per cent. of raw sugar, in color equal to fair refining, but intrinsically much superior; or it would yield 5½ per cent. of sugar equal in every respect to New York refined "B."

The beets raised by the Messrs. Gennert, if successfully and rapidly worked up, would have produced not less than 450,000 pounds of refined sugar.

I learned from Mr. Gennert the following particulars of the difficultities they have met with in carrying on their operations since harvest:

In the first place, their machinery, instead of being completed during the summer, so as to be ready for work by the time when the beets were ripe (in September), was only got in starting order by the 5th of December last. They then commenced operations with green inexperienced hands, and during the first few days made very slow progress, notwithstanding the beets were found to work and yield well.

After five days of work, December 10th, the vacuum-pan collapsed, which misfortune entirely stopped all work until a new pan could be obtained. This was ready with the beginning of the new year, when they commenced again, only to meet with new discouragements. First, the supply of water proved to be inadequate to their requirements, and steps were immediately taken to deepen the wells, so as to reach below the hard pan, and they expect now to have obtained a full supply of water. The next, and to a sugar manufacturer the most serious difficulty of all, was a too limited supply of steam, which they were trying to remedy at the time I was there. They have depended on five two-flue boilers, which were not well set, the smoke being carried through a narrow breach flue into a narrow and low sheet-iron smokestack; the entire arrangement being not well adapted for the proper consumption of the Lasalle or Fairbury coal. The workmen about the place seemed to think the insufficient supply of steam the main drawback to success. In all other respects the works appear to be well appointed. They are built to run on the centrifugal system, and are provided with clarifiers, scum-presses, bone-black filters, retorts, vacuum-pan, and such other machinery as is generally found in a manufactory of this class.

All machinery is of modern construction and well adapted to the work. The capacity of the manufactory is estimated to be equal to 50 tons of beets per day. During the few days the works have been in operation they have turned out about eighteen thousand pounds of sugar (two thirds of which was equal to N. Y. Ref. B.), which was the product of an unknown quantity of beet roots, as I found they had not kept any record of the quantity brought from the pits to the factory. The pulp was not watered in the centrifugals, so as to save evaporation. The juice was boiled blank, and placed in large tanks to crystallize. This course was mainly taken to economize the use of steam. The first product granulated in twenty-four hours and the second in three days, so as to go in centrifugal machines. I could see nothing of the third product. I very much regret that the Messrs. Gennert could not give me an accurate account of the cost of cultivating their beets; the estimate, as I have already said, was less than four dollars per ton. It is also to be regretted that no account of the weight of beets taken to the factory was kept, although any calculation made on that basis would be unfair, considering the irregular operations of the factory and the deterioration in the saccharine properties of the roots from long delay in working.

Notwithstanding all the difficulties the Messrs. Gennert have encountered (most of which, however, were avoidable and should have been foreseen), I found that they manifest no feeling of discouragement, and I fully anticipate their success another season. The result of my investigations, added to my previous knowledge of the subject, more than ever confirms my belief in the speedy and successful development of this branch of agricultural industry. And this feeling is already widely entertained through the West, where suitable lands and abundance of fuel can be had at low prices, in the immediate vicinity of a ready market for all products of the manufactory.

In conclusion I will say, that I know of nothing to prevent an individual or company, possessed of a knowledge of the subject, making this a business of large profits from the beginning.

Very truly yours,

(Signed,) R. W. Bender.

New York, Feb. 12, 1867.

I also annex extracts from a letter written by Charles Belcher, President of the Belcher Sugar Refining Company of St. Louis.

* * * * *

To-day Mr. Holm had a short letter from Theodore Gennert, who returned a fortnight since. He states that they are still making sugar, and with satisfactory results, and will continue to work their beets until it becomes necessary to look after out of door work. He speaks well of the quality of the juice.

* * * * *

From all I know or have heard, I would suppose $4 per ton, $40 per acre, a very liberal estimate for the cost of raising and harvesting beets. Gennert told me he kept a pretty accurate account of the cost of raising his crop of beets two years since, and that $3 per ton would cover it. I would think also that if properly prepared for the work, with suitable buildings, machinery, apparatus, and fixtures, and with well-informed and judicious management, the cost of making sugar from the beets should not exceed $4 per ton of beets.

I cannot see any good reason why, with the right preparations and good management, about as good results cannot be obtained from beet sugar making in this country as in Europe.

* * * * *

Mr. Gennert was not properly prepared for working his beets advantageously and profitably; he has made mistakes in his calculations and arrangements, and his business this year will not be a success; but he seems to have shown us that beets can be raised in sufficient quantity, and we have evidence that they have sugar in them that would pay well for the working in Europe.

I have felt that it would be a great advantage to this country if the manufacture of beet sugar could be successfully introduced, and we have assisted Gennert's enterprise by loaning him machinery and subscribing to his company, to aid in developing it; and I have all along had strong faith in its practicability, and still have it.

* * * * *

Yours truly,

(Signed,) CHARLES BELCHER.

ST. LOUIS, March 19, 1867.

I also annex an extract from a letter written by Mr. Bender on the subject: —

NEW YORK, March 10, 1867.

* * * * *

Ere long I expect to see general attention directed to this industry. An impetus once, and properly given, must develop the manufacture of domestic sugar rapidly, spreading its benefits all over the land, enriching the farmer and the mechanic, opening new channels of support to thousands, stimulating good husbandry and inventive genius. As early as 1861 I felt convinced that the "sorghum cane," from the chemical nature of its juice, and from the difficulty of bringing it to maturity, would prove a failure as a sugar producer, and only a partial success as a syrup producer, and that for the range of the Northern States, the sugar beet only possessed all the qualifications for extensive and reliable production of sugar in an eminent degree. I induced, at that time, William H. Belcher, of the St. Louis and Chicago refineries, to import some beet seed from Europe, which, for experiment, we distributed amongst farmers in the West, and the results were of the most encouraging character. Further investigations by us and others in the following years satisfied us that the yield per acre was 10 to 20 tons, at a cost of less than $4 per ton; that the saccharine qualities of the American beet are equal to those of the European, and that there is no more difficulty in making sugar in America from beets than there is in Europe; and further, that from sugar-beet a good merchantable raw sugar can be produced at a cost of less than *five* cents per pound, such sugar being worth in New York to-day, taking the color as a standard, 10 cents; but intrinsically, its value would be much greater; the smaller admixture of grape sugar warranting to the refiner a larger yield of refined sugar, and less in syrup.

By the death of Mr. Belcher, two years since, this industry lost one of its warmest supporters. For my own part, I never had sufficient liberty of action to engage myself in this business, although I have constantly endeavored to interest people in a matter which I consider of the utmost importance to this nation.

(Signed,) R. W. BENDER.

I also annex extracts from a letter written by A. J. Corning, a chemist and sugar refiner, who has been for several years engaged in the Adams Sugar Refinery of Boston:

NEW YORK, March 15, 1866.

* * * * *

All my inquiries tend to convince me that beets can be raised under $4 per ton, and from my experience in refineries I should say that the sugar can be extracted easily at 3 cents per pound. Assuming the yield of sugar to be 6 per cent., which is much below the usual yield (a German who has had charge of a manufactory in Germany told me that he obtained 9¼ %, while a neighbor obtained 9¾ %), we would have the following as the cost per pound of sugar:

2240 lbs. beets	$4.00
6 % sugar = 134 lbs. @ 3 cents	4.02
Total cost per ton	$8.02
134 lbs. sugar @ 6 cents	8.04

Giving the cost of producing at 6 cents per lb. for sugar a great deal better for refining than the sugar we import for that purpose at from 9 to 12 cents per lb. The sugar is what refiners term "strong" (containing very little foreign matter), and yields a larger percentage of refined sugar than that of Cuba.

The sugar of the beet is identical with that of the cane, and possesses the same sweetening power.

The manufacture of beet sugar in this country is deserving of the highest consideration, both as regards the development of our agricultural, manufacturing, and commercial interests.

(Signed,) A. J. CORNING.

Mr. Bender and Mr. Corning are both confident that an abundant supply of beets can be obtained by manufacturers at from $3 to $4 per ton; that they can be manufactured into sugar at from $3.50 to $4 per ton; and that they will certainly yield 6 % of white sugar, worth in Chicago at least fourteen cents per pound.

Assuming that their figures are correct, of which there is not a particle of doubt, and taking their highest estimates as the basis for a calculation, the following result could be produced by a company with $350,000 capital, which is sufficient to erect a mill of a capacity to work 30,000 tons of beets each season.

30,000 tons of beets, costing $4 per ton . . .	$120,000
30,000 " " to work $4 " . . .	120,000
	$240,000
Producing	
1,800 tons sugar (being 6 %) at 14 cents per lb., or $280 per ton	$504,000
6,000 tons pulp @ $2	12,000
900 " molasses @ $20	18,000
	$534,000
Less expenses	240,000
Profit (being 84 % on capital), ...	294,000

In the pamphlet of Mr. Walsh, to which reference has been made, he says, —

"The introduction of beet sugar as a staple product of the United States, but especially for the vast fertile prairies of the West, has claimed the profound attention of statesmen and eminent practical citizens for more than a quarter of a century. Much has been written upon it; much information has been diffused; many interesting and thorough experiments have been made, and the general results have been in the highest degree satisfactory. All the inquiries and investigations that have been made, and facts gathered, strengthen the conviction that it is only necessary to engage systematically in the culture of the root and the manufacture of sugar in the United States, to insure results of the highest national importance, and establish in our borders a tillage that will improve our system of husbandry, an employment that will give a wider scope to our energy and industry, and a manufacture that will supply us in abundance with a great staple of consumption, for the largest portion of which we are now dependent upon other countries to supply.

"In 1837–8 Henry Clay was actively interested in the question of beet sugar as a crop for the United States. He had watched the rise and progress of this industry in France: had made himself familiar with the details of the culture of the root, the manner of extracting its sugar, the success attending it as an economical measure, and his sagacious mind grasped at once the full importance of this grand national resource for the great West, which he loved so well. He made it a topic of his letters; he introduced it in his speeches; his conversation abounded with allusions to it; and he has left on record full evidence of the constant faith he had that the West would some day be as famous for its production of sugar as it has become for the production of the cereals. The granary of the world, it may also be the sugar-grower for the world. It is the home of intelligence, and industry, and enterprise; and these forces, united to the exhaustless producing capacity of the soil, will secure success in whatever undertaking her farmers and her men of activity may engage.

"But it was not Mr. Clay alone whose inquiries kept pace with the progress and improvements in the beet sugar manufacture, and who, with a full knowledge of the qualities of the root, the nature of our climate, and the capacities of the soil, had full faith in the peculiar adaptation of this culture and manufacture to a very large portion of the Uuited States.

"From time to time fields have been cultivated with beet root, for the express purpose of sugar-making; and in numerous instances that could be cited, — that are indeed recorded in the pages of agricultural journals of the day, — the success of the efforts and experiments has proved fully commensurate with the expectations that had been entertained. Sugar of the first quality has been produced, and with no more trouble than is experienced in the manufacture of maple sugar. These processes have been carried on in Maine, in Massachusetts, in New York, in Pennsylvania, in Ohio, and in Illinois, with equal success.

"But, though the experiments alluded to amounted to actual demonstration, no experiment or enterprise on a large scale, with machinery or equipments to manufacture sugar in large quantities, was attempted. Indeed the prevailing idea seemed to be, at that time, to introduce beet-sugar as a domestic manufacture, and efforts were directed rather to the introduction of processes and machinery whereby each farmer might make his own sugar, than to the establishment of the business on a large scale. If this was an error, as we are disposed to think it was, no great harm arose from it, inasmuch as, from the prevailing low price of sugars, interest in the subject subsided, and finally it was laid aside altogether.

"The truth is, the country was not ripe for the enterprise. It did not feel the need of emancipating itself from dependence upon the sugar of other nations. The people were recovering from the great commercial disaster of 1837, and new enterprises and new speculations that promised readier and greater returns engaged their energies.

"Besides, the country was not ripe for it in another respect. We were less than twenty millions of people. The emigration to the West was draining the young and active elements of the population from the old States, and the emigrants were more intent upon establishing their domiciles in new locations than upon engaging in new manufactures. Those great States, which have outstripped in population, wealth, and influence so many of the older States, were then settling; their character had not been determined; their resources had not been ascertained; their great and glorious future had not been revealed.

"We are now in a changed condition of things; our twenty

millions have increased to more than thirty millions, and the ratio of increase is undiminished. The West has become settled. It wants and must have new staples from the teeming soil, new employments for labor, and cannot afford to let any source of wealth go unimproved.

"Besides, the war has disturbed the routine into which we had fallen, imposed new burdens and new duties upon us, and has pointed significantly to the controlling duty which devolves upon us to strive to render the country independent of foreign supplies for those great articles of consumption which form the necessities of life. If the war has taxed our means, it has aroused our energies; if it has disturbed our peaceful pursuits, it has developed our strength; if it has tried our system of government, it has shown us that we have unequalled elements of greatness.

"The exigencies of the war have aroused the nation to an attentive consideration of everything that will make for the common advantage; and new employments are opened before us for our abundant land, capital, and men. Situated midway between the continents of the old world, in temperate latitudes, with every variety of soil, and land sufficient for the sustenance of one half the estimated population of the globe, there is nothing to hinder us from producing, in the most profuse abundance, everything which is produced in the same zone in other lands."

www.ingramcontent.com/pod-product-compliance
Lightning Source LLC
LaVergne TN
LVHW021405110826
845150LV00007B/1793
9781425511975